IMAGES OF WAR

THE LITTLE GUN THAT LED THE BLITZKRIEG

AF813521

IMAGES OF WAR

THE LITTLE GUN THAT LED THE BLITZKRIEG

THE RISE OF THE GERMAN 3.7CM ANTI-TANK GUN

JACEK ZABIELSKI

Pen & Sword
MILITARY
AN IMPRINT OF PEN & SWORD BOOKS LTD.
YORKSHIRE – PHILADELPHIA

First published in Great Britain in 2026 by
Pen & Sword Military
An imprint of
Pen & Sword Books Ltd
Yorkshire – Philadelphia

Copyright © Jacek Zabielski, 2026

Print ISBN 978 1 03619 635 6

The right of Jacek Zabielski to be identified as the Author of this work has been asserted by him in accordance with the Copyright, Designs and Patents Act 1988.

A CIP catalogue record for this book is available from the British Library.

All rights reserved. No part of this book may be reproduced, transmitted, downloaded, decompiled or reverse engineered in any form or by any means, electronic or mechanical including photocopying, recording or by any information storage and retrieval system, without permission from the Publisher in writing. NO AI TRAINING: Without in any way limiting the Author's and Publisher's exclusive rights under copyright, any use of this publication to "train" generative artificial intelligence (AI) technologies to generate text is expressly prohibited. The Author and Publisher reserve all rights to license uses of this work for generative AI training and development of machine learning language models.

Typeset in 12/14 Gill Sans by
Typeset by SJmagic DESIGN SERVICES, India.
Printed and bound in England by CPI Group (UK) Ltd., Croydon.

The Publisher's authorised representative in the EU for product safety is
Authorised Rep Compliance Ltd., Ground Floor, 71 Lower Baggot Street,
Dublin D02 P593, Ireland.
www.arccompliance.com

For a complete list of Pen & Sword titles please contact

PEN & SWORD BOOKS LIMITED
George House, Units 12 & 13, Beevor Street, Off Pontefract Road,
Barnsley, South Yorkshire, S71 1HN, England
E-mail: enquiries@pen-and-sword.co.uk
Website: www.pen-and-sword.co.uk

or

PEN AND SWORD BOOKS
1950 Lawrence Rd, Havertown, PA 19083, USA
E-mail: uspen-and-sword@casematepublishers.com
Website: www.penandswordbooks.com

MIX
Paper | Supporting responsible forestry
FSC® C013604

Contents

Introduction

At first glance, the modest 3.7cm *Panzerabwehrkanone* (*Pak*) may appear underwhelming when measured against the monstrous tanks and heavy artillery that came to define the Second World War. Yet this seemingly small-calibre gun had an outsized influence on early German military success and deserves its rightful place in the annals of twentieth-century military history. Conceived in the late 1920s and introduced into mass production in the early 1930s, the 3.7cm *Pak* was the cornerstone of Germany's anti-tank defence doctrine throughout the interwar years and the opening stages of the Second World War.

Deployed across nearly every branch of the *Wehrmacht* – from the infantry to airborne troops, and especially within panzer divisions – this gun's remarkable mobility and adaptability made it ideal for the lightning-fast tactics of *Blitzkrieg* ('Lightning war'). Its light weight, compact form and motorized towing capability allowed it to accompany advancing columns, provide swift battlefield deployment and offer infantry formations a credible answer to enemy armour.

By the time war broke out in 1939, the 3.7cm*Pak* had already been produced in the tens of thousands, benefiting from standardized design, low manufacturing cost and ease of maintenance and repair, as well as the supply of ammunition. It became one of the most mass-produced anti-tank guns in the history of the German Army. Although technological advances in tank armour would soon render it increasingly obsolete, particularly against heavily protected Allied and Soviet vehicles, its low silhouette and agility continued to lend it value in ambush tactics and supporting roles.

Throughout the war, the gun was seen in action on every front, from Norway's frozen fjords to the deserts of North Africa, the muddy expanses of Eastern Europe and the fields of France. Whether firing from roadside ditches in Belarus or mounted on Panzer IIIs or self-propelled platforms racing across the steppes, the 3.7cm *Pak* embodied the early-war ethos of German military engineering: efficient, adaptable and deadly – until it no longer was.

This monograph aims to trace the development, deployment, variants, tactical use and ultimate legacy of the 3.7cm *Pak* gun. Through archival photographs, technical data, first-hand accounts and contextual analysis, it uncovers the story of a weapon that, while small in stature, played a giant role in shaping the early armoured warfare landscape of the Second World War.

A Note on Terminology

Early German anti-tank guns developed during the mid-war period by the *Reichswehr* were originally designated as *Tankabwehrkanone* (abbreviated *Tak*). A new official designation for anti-tank guns was introduced on 7 July 1936 (per *Allgemeine Heeresmitteilung*), changing the term to *Panzerabwehrkanone* (abbreviated *Pak*).

This designation was revised again in 1941, when the term *Panzerjägerkanone* (also abbreviated *Pak*) was adopted, according to *Allgemeine Heeresmitteilung* dated 24 April that year.

In official German nomenclature, the gun calibre preceded the designation. If the calibre was greater than 20 millimetres, it was expressed in centimetres (e.g., 37mm = 3.7cm). Therefore, a 37mm anti-tank gun would be officially referred to as a 3.7cm *Pak*.

However, in many post-war publications, the designation *Pak 35/36* became popularized, despite it not strictly aligning with official German terminology.

German unit names are presented in full upon first mention, in their original form (e.g., *Panzer Abwehr Abteilung* = Anti-Tank/Tank Hunter Battalion), with abbreviations such as *Pz.Abw.Abt.* used thereafter. The commonly understood term 'Panzer' is retained throughout due to its widespread recognition among English-language readers. Similarly, German military notation conventions are followed, such as using periods in ordinal numbers (*6. Pz.Div.* for 6th Panzer Division), distinguishing unit sizes with Roman numerals for battalions (e.g. *I./Pz.Rgt. 11* = 1st Battalion/11th Armoured Regiment) and using Arabic numerals for companies (*Kp. 1*, *Kp. 2*, etc.). The term *Abteilung* (*Abt.*) is translated as 'battalion' when it refers to operational formations, and as 'detachment' when referring to non-combat or administrative elements.

Chapter One

Forging the *Pak*

During the final stages of the Great War on the Western Front, the German Army was compelled to develop a new type of gun in response to increasingly frequent massed Allied tank assaults. Rather than creating a weapon from scratch, the decision was made to adapt an existing trench gun (*Schutzengrabenkanone*), specifically the 3.7cm *Revolverkanone* (3.7cm *Rek*), a French Hotchkiss design produced under license in Germany.

Originally mounted on a fixed base, the gun required significant modification to serve as an effective anti-tank weapon. This included redesigning the lower carriage and mounting it on a wheeled chassis to improve battlefield mobility. Though not designed for towing, the modified gun could be manually manoeuvred by its crew, hence the use of a simple chassis with metal-spoked wheels – basic but functional for the time.

To suit the anti-tank role, the gun's original triple-barrel configuration was reduced to a single barrel, as high-volume fire was unnecessary against armoured targets. Initially, the *Rek* gun fired only 37×94R high-explosive fragmentation rounds, prompting the development of new armour-piercing projectiles, both with and without tracer capability.

The redesign effort was led by *Oberstleutnant* Karl Fischer of the Artillery Testing Commission (*Artillerie-Prüfungs-Kommission, APK*), and the weapon was subsequently named the 3.7cm *Fischer Tankabwehrkanone* (3.7cm *Tak*), incorporating the designer's name.

In August 1918, the *APK* ordered a limited production run of fifteen Fischer guns from the *Artillerie-Werkstatt* (Artillery Workshop) in Munich, with delivery scheduled within two months. The gun underwent field testing at the Meppen proving ground in October 1918. However, due to the armistice in November, the weapon was never deployed in combat.

Separately from Fischer's project, the Imperial Army High Command initiated parallel developments in August 1918, commissioning Rheinmetall (Düsseldorf) and Friedrich Krupp AG (Essen) to design similar anti-tank guns. The prototypes were mounted on the carriage of the 7.58cm *Leichter Minenwerferneuer* (light trench mortar), though they differed in chassis configuration. Krupp's gun featured an adapted trench mortar base without wheels (*ohne Radem*), whereas Rheinmetall's version utilized the trench mortar's base and chassis with wooden wheels. As additional types of backup anti-tank guns, there were 20mm guns designed by Becker and Erhardt (Rheinmetall). Rheinmetall completed the first prototype of its gun on 29 October 1918. This semi-automatic gun had a barrel length of 45 calibres and fired armour-piercing projectiles weighing 750g with an initial velocity of 690m/s, a recoil of 300mm, and a firing line height of only 590mm. The weight of the gun ready for firing (in the firing position) with twenty rounds was 250kg.

The anti-tank guns were intended for allocation to battalions of trench mortar units, each to be equipped with thirty-two guns. These units functioned as HQ strategic reserve/rapid-reaction forces.

The initial order for 3.7cm anti-tank guns (*Tankabwehrkanone*) amounted to 300 units by Rheinmetall and 250 guns by Krupp. However, Krupp failed to start production (likely producing only one gun in 1918) before the war ended. Rheinmetall, after the war's conclusion around 1918–1919, delivered nearly 600 guns, but they had to be destroyed under agreements with the victorious powers.

The 3.7cm *Tankabwehrkanone* (Fischer system) was transported to its position on a simple cart and then lowered onto a tripod mount. The artillery workshop in Munich produced fifteen of these guns before the end of the war. (Bayerische Armee Museum in Ingolstadt)

(**Above**) Rheinmetall also produced the 3.7cm *Tankabwehrkanone* which was fitted with a simple ring-type gunsight and large gunner saddle, while the ammunition was stowed in two boxes mounted on the front of the carriage (Museum in Koblenz)

(**Opposite, above**) A 3.7cm *Tak* (Rheinmetall model) half-battery is seen here on manoeuvres in the summer of 1918. Each gun was manned by a crew of five soldiers – exactly the same number later assigned to the improved 3.7cm *Pak* developed in the 1930s – comprising a gun commander, layer, loader and two ammunition handlers or assistants.

(**Opposite, below**) The 3.7cm *Tankabwehrkanone* is shown in the ready-to-fire position, manned by the gunner and his spotter. Aiming was straightforward, relying on enlarged, rifle-style iron sights – essentially a scaled-up version of the front post and rear notch. No optical sights were mounted, as target ranges for enemy tanks were expected to be short and engagements quick. The spotter's role was crucial, helping the gunner track moving targets, judge distance and relay corrections, especially when firing under battlefield stress.

(**Above and below**) Improvised 'KD I Type 2' armoured cars, armed with 3.7cm Fischer guns, were employed by the *Freikorps von Epp* during operations in Germany against communist forces in the turbulent years of 1919–1920. Armoured cars were intended for urban fighting and convoy protection, where mobility and intimidation mattered as much as firepower. Large placards fixed to the hulls warned civilians: 'Keep clear of the combat vehicle area, or *we will open fire*,' underscoring their role in crowd control and street engagements.

Chapter Two

Designing a Tank Killer

Representatives of the German government, signing the peace treaty with the victorious Entente in Versailles near Paris on 28 June 1919, agreed to a series of political, economic and military restrictions. These provisions significantly curtailed the size and firepower of the German Army, including restrictions on tanks and armoured vehicles.

Germany was permitted to retain only 1,000 anti-tank rifles (*Panzerschreck*), a number immediately considered inadequate by the *Reichswehr*. While the development of armoured vehicles was halted, Germany was allowed to pursue research in anti-tank defence. It is also possible that Rheinmetall managed to transfer some documentation and several 3.7cm *Tak* guns to the Netherlands before the Treaty of Versailles restrictions took effect.

Fortunately for Germany, the treaty allowed specific industrial facilities to produce limited quantities of weapons and ammunition necessary for training and replacing worn-out equipment of the *Reichswehr*. Consequently, the Interallied Military Control Commission permitted leading German firms in artillery production, such as Krupp and Rheinmetall, to maintain their design offices. Both companies were officially authorized to retain design and production documentation for various types of weapons, ensuring continuity in their engineering capabilities. Many highly skilled designers, engineers and technicians were able to remain employed in these facilities.

Clandestine efforts to develop a new type of anti-tank gun were only initiated in mid-1924. Earlier attempts were hindered by the unstable political and economic conditions in Germany, exacerbated by the occupation of the Ruhr Valley by French and Belgian troops from 11 January 1923 to 25 August 1925.

The German armament factories of Krupp AG in Essen were situated in the occupied territories, and although formally not supposed to be occupied, they fell under French control.

On 26 January 1925, a conference convened in Berlin under the leadership of Major General Friedrich von Haack, Chief of Staff of the Army

Command (*Heeresleitung*) of the *Reichswehr* Ministry, with representatives from the artillery inspectorate and *Technisches Amt und der Waffenamt* (protocol Number 18/25 Jn IIIA dated 3 February 1925), focusing on anti-tank guns.

Following the conference, it was decided to develop a light infantry gun (later known as the *le.I.G. 18*) and 37mm anti-tank gun (the future 3.7cm *Pak*). The design and development of mounts for both systems commenced almost immediately. Initially, the plan was for both guns to share the same mount, but the final designs turned out to be significantly different. The only shared components were the wheels, which were initially wooden but were later replaced with metallic ones. Eventually, the suspension springs were also standardized.

Foreign occupation forces withdrew from the Ruhr in August 1925, and by February 1926, the *Inspektion für Waffen und Gerät zum Heereswaffenamt (HWA)*, later known as *Waffen-Prüfung 4 (Wa-Prüf 4)*, issued official requirements for the development of such a gun.

The technical guidelines issued on 3 February 1926 established the following tactical and technical parameters:

The anti-tank gun was to be a wheeled weapon designed for front-line operations, capable of being moved by its crew using transport straps (referred to as 'draught traces') or towed by a single horse. The gun's barrel was to have an elevation range from -5° to +15°, and a horizontal traverse of 20° to either side. At that time, no specific requirement was set for the minimum height of the line of fire. The wheelbase measured 1,125mm.

A 6mm thick armour shield was included to provide protection against 7.92mm rifle rounds at a distance of 500 metres. The weapon's initial design goal was to keep its weight under 250kg; however, this limit was ultimately exceeded during development.

The armour-piercing projectiles were to be developed to penetrate 40mm nickel steel armour at a 60° angle from a minimum distance of 500m. Upon penetration, the projectile was expected to generate sufficient fragments to incapacitate the crew inside an armoured vehicle. Additionally, the gun was expected to engage targets such as machine guns protected by frontal armour (like the Russian/Soviet Maxim model 1905/1910) or light field fortifications. For this purpose, alongside armour-piercing projectiles, the gun could also fire high-explosive fragmentation shells.

The requirements also specified that the carriage of the anti-tank gun should feature double, folding trails to ensure stability during firing at low angles. With the trails spread at 60°, the horizontal firing angle was

40°. However, when necessary, firing could be conducted with the trails closed, reducing the firing angle to just 6°.

It was proposed to utilize an already existing barrel from the 3.7cm *Tak 18* gun, which had a length of 45 calibres and featured a semi-automatic breech similar to the one used in the Fischer system gun.

The task of designing the new gun was assigned to the design office of Rheinnische Metallwaren- und Maschinenfabrik (Rheinmetall).

On 1 July 1925, the Army Weapons Testing Office (*Heereswaffenamt Prüfwesen*) formulated a priority list of projects, with the development of the new 3.7cm *Tak* gun being prominently featured. The responsibilities encompassed not only designing and prototyping the gun but also manufacturing suitable ammunition, a tractor (*Protze*) and an ammunition carrier (*Munitionswagen*).

The design and refinement of the gun spanned over two years. In the subsequent urgent project list ('urgency list for *Wa Prw.*') dated 1 January 1927, the production of the 3.7cm *Tak* gun was given top priority. Furthermore, there was a request to initiate the production of the 3.7cm *Panzersprenggranate mit Lichtspur* or high-explosive shell with a base fuse (*Bodenzunder*) and the 3.7cm *Messinghulse*, which refers to the brass cartridge casing for the shell.

On 1 August 1927, a report was issued detailing the current status of the defence industry and the progress of prospective projects towards production readiness. In the section labelled 'current status/detailed description', specific information was provided regarding the *le.I.G. 18* and 3.7cm *Pak*:

> 'Preparations for mass production of both guns could potentially be completed by 1929 at the earliest. It remains uncertain whether the new 37mm anti-tank guns will be ready earlier. Two experimental prototypes of the 3.7cm Tak are slated for completion by January 1, 1928. In case of urgent need, final blueprints could be prepared before the beginning of 1929 if trials, including field trials, are expedited, allowing for preliminary production to commence. Quality assurance measurements and technical specifications could be finalized before the end of 1929, paving the way for mass production. Fourteen light infantry mod. 1918 guns (second experimental prototype) will be ready by January 1, 1928, and will undergo field trials.'

A meeting held on 9 November 1927, which included the Chief of the Administrative-Quartermaster Directorate, the Chief of the Armament Directorate and directors from various inspectorates of the Armament

Directorate, led to the composition of a memorandum by the infantry inspectorate. This memo highlighted that the final design of the 3.7cm *Pak* had not yet been settled:

> 'There are disagreements regarding the future use of the 37mm gun. The technical and administrative-quartermaster directorates insist on mounting it on a self-propelled platform for mobile anti-tank defence. The infantry inspectorate prioritizes a gun towed by horse or tractor, in addition to a self-propelled option.'

Consequently, the decision was made to first evaluate the characteristics of the 37mm barrel. Detailed discussions on mobility issues would follow later. Ultimately, the *Rheinmetall-Waffe* variant with split trails and wooden wheels for horse-drawn transport was chosen.

The gun underwent two years of development and refinement. Trials began in January 1928 at the Kummersdorf proving ground. Following successful testing, it was officially adopted into service as the 3.7cm *Tak* (*Tankabwehrkanone*) – *Geheime Kommandosache No. 6163/32*, dated 20 November 1932.

It is important to note that this version of the gun was designed to be towed by horses, featuring unsprung wooden wheels with steel rims, which significantly restricted the marching speed of the gun crew to just a few kilometres per hour.

By May 1929, a batch of thirteen guns was ready, including five allocated for trials conducted by the *Heereswaffenamt* at the Kummersdorf proving ground. The first guns from serial production became available in January 1930.

Earlier, on 24 May 1929, funds totalling 6.2 million RM were reserved under the *Notrustung im Kriegsfall* (Armament in Case of War) for the purchase of 577 x 3.7cm *Tankabwehrkanone* guns with *Spreizlafette* mounts. According to report 6163/32, dated 20 November 1932, the peacetime requirement for the *Reichswehr* was set at 606 guns. At that time, the land army possessed 201 guns, with sixty-three in reserve, resulting in a shortfall of 344 guns needed to meet the planned establishment.

In the early 1930s, manoeuvres involving the *Tak* guns highlighted the necessity of concentrating anti-tank artillery along the axis of tank assaults to maximize effectiveness. This realization underscored the need for anti-tank artillery units to match the mobility of tanks themselves, necessitating their motorization.

By order issued on 17 September 1934, the standard 3.7cm *Tak* gun underwent modernization. It was outfitted with wheels featuring pneumatic tyres (typically tube-type or occasionally solid rubber tyres) and an updated suspension system, allowing it to be towed by trucks and specialized tractors. The lower carriage was also redesigned to increase the horizontal firing angle to nearly 60°, and minor adjustments were made to the frontal armour of the gun.

Interestingly, by November 1934, the first prototype of the motorized traction gun was ready for testing.

Subsequent minor design refinements in 1935 and 1936 included advancements in barrel manufacturing technology, improvements in wheels and towing mechanisms, and the introduction of a new telescopic sight (incorporating panoramic, telescopic and collimating sights). As a result of these developments, the gun was designated as the 3.7cm *Panzerabwehrkanone 35/36*.

In addition to the standard 3.7cm *Pak* variant, a lightweight version was developed for equipping airborne units. This variant featured lighter carriage trails, a thinner frontal armour plate and other components reduced in weight. Known as the 3.7cm *Pak mit Leicht Lafette fur Fallschirmjäger*, externally designated as the 3.7cm *Pak 35/36 auf Leichter Feldlafette*, it closely resembled the standard 3.7cm *Pak*.

The gun was specifically designed for airborne transport, mounted externally on the fuselage and undercarriage of the Ju-52/3m transport aircraft. During flight, special wooden blocks secured the gun components. Two ammunition boxes were fixed to the carriage trails, with additional ammunition stored in containers that could be airdropped from Junkers aircraft. A package containing five standard parachutes used by German paratroopers was mounted between the carriage trails. Upon landing, the gun utilized wheels theoretically designed to absorb ground impact, necessitating reinforced chassis elements.

Initially deployed with the 7th Parachute Anti-Tank Battalion (*Fallschirm Panzerjäger Abt.* 7) of the 7th Parachute Division (*7. Fallschirmjäger Div.*), and later distributed in smaller quantities to other parachute units, approximately 100 guns were modified in this configuration.

During wartime production, further enhancements were made, including the introduction of a monobloc barrel, redesigns to the breech mechanism and chassis reinforcement to facilitate higher towing speeds.

Design team with the prototype of the 3.7cm *Tak L/45*, led by Dr Waninger, produced a remarkably advanced design for its time.

Reichswehr officers inspect the gun drill of a 3.7cm *Tak* gun crew from a cavalry regiment. Note the M.1919 cavalry helmets, with ear cut-outs, used only by the horse artillerymen.

The 3.7cm *Tak* could be man-hauled by its detachment, when necessary, due to its size and weights. Two leather bandoliers, hauling and lifting the gun, were issued to each *Tak* crew.

A *Reichswehr* officer addresses a gathered crowd, flanked by artillery pieces (15cm *schwere Feldhaubitze 13 (15cm sFH13)* produced from 1913–1918 and brand new 3.7cm *Tak* gun). Such public events were common in the interwar period, often staged to inspire patriotism, showcase military strength or remind of past triumphs.

A commemorative photograph of *Reichswehr* soldiers proudly presenting their newly issued *Tak* guns – the very first modern weapons developed after 1918, despite the severe restrictions imposed by the Treaty of Versailles.

The first *Tak* guns had a long barrel, wooden wheels and no suspension springs, as they were designed for being towed by horses.

Reichsfuhrer Himmler inspecting gun drill of the *3.7 Tak* gun crew in *SS-Freiwillingen Gebirgs Div. 'Prinz Eugen'* in Croatia on 7 December 1943.

Initially, the *Pak* section in *Reischwere* infantry division consisted of two cross-country towing cars, with the first towing the gun and the second towing an ammunition trailer.

An anti-tank section equipped with *3.7 cm Pak* guns from the *Kavallerie Rgt.18* "defending" against a bicycle platoon "attacked" by a Kfz 13 armored car. This photo was taken near Bad Cannstatt during field exercises in 1936.

A *Pak* crew takes a short break during pre-war field exercises. Each soldier wears a red elastic band around his helmet, a simple yet effective marker identifying them as part of one of the exercise's opposing forces. These reversible bands – most commonly red and yellow, and sometimes red and blue – allowed for quick re-designation, ensuring that both sides in the drill could be distinguished at a glance.

After completing field training, gun crews were required to meticulously clean their equipment, washing away mud and dust before returning to barracks. Streams and ponds in the field often served as their only source of water for this task.

Time in garrison was devoted to perfecting maintenance routines, mastering muster procedures, studying regulations, and even staging public demonstrations – meant to inspire confidence and bolster morale among civilians

(**Above**) *Kübelwagens* (utility cars with a rugged, open design) were employed by the *Reichswehr* as staff cars, light towing vehicles for anti-tank and infantry guns, and a specialized platform for signal and engineering units.

(**Opposite, above**) After winter exercises in 1936, a *Pz.Abw.Abt.* of an unidentified infantry division parades through a village in the Harz Mountains.

(**Opposite, below**) Crews of *Pak* guns from an unidentified *Wehrmacht* anti-tank battalion pose for the camera in 1936. The guns still display the standard *Reichswehr* three-tone camouflage pattern, consisting of RAL colours No. 17 *Erdgelb* (RAL 8002/RAL 0704030) or No. 22 *Honiggelb* (RAL 1005), combined with No. 18 *Braun* (RAL 8010) or *Sepiabraun* (RAL 8014), and No. 27 *Grün* (RAL 6002) or No. 28 Grün (RAL 6007). This camouflage scheme was applied to *Pak* guns and their motorized tractors from 1933 until it was phased out in 1937.

P
WH 93041

(**Above**) After firing, carbon deposits and residue had to be scrubbed out using brushes, oil and rags. Because the bore was small, fouling could build up quickly. Cleaning soldiers are wearing the white (or off-white) *Drillichanzug*, a fatigue uniform, which absorbed dirt and sweat, leaving the service uniform clean for inspection or combat. Since these were worn on training grounds, camouflage was not a concern. In fact, their plain light colour made it easy for instructors to spot trainees in exercises

(**Opposite, above**) A photograph depicting the *Sf.14Z* scissor binoculars, a vital optical instrument used by officers overseeing gun-firing drills. With its distinctive silhouette, the device quickly earned the nickname 'donkey ears'.

(**Opposite, below**) A group of German soldiers proudly pose with their 3.7cm *Pak*. Many soldiers kept such photographs as personal mementos to send back home to their families.

Pre-war field training. Note the wicker ammunition container, occasionally used during exercises. Such woven carriers were lightweight, durable, shock-absorbing and inexpensive to produce, making them easier to transport by hand across rough terrain.

From 1940 onward, the 'white' *Drillichanzu* were gradually replaced by reed-green or field-grey fatigue uniforms (still called *Drillich*), but the white versions remained in circulation, especially in training anti-tank units and among recruits

An unidentified anti-tank crew is seen here during pre-war winter exercises. The exact battalion cannot be determined, but some details offer important clues. Most telling is the tactical number '24' painted on the ammunition box. German practice at the time assigned sequential numbers to companies. In this case, numbers 1–12 were allocated to the 1st Company, 13–24 to the 2nd Company and 25–36 to the 3rd Company.

Pak guns on the move … while the vehicles are already registered to the *Wehrmacht* (WH-29611, WH-29778), the guns are still painted in the *Reichswehr* tri-colour disruptive camouflage.

(**Above**) Manhandling the gun across flat terrain demanded the strength of four men, but dragging it up even a modest hill required twice the manpower and relentless effort. The strain was brutal, yet such manoeuvres were practiced again and again – because in battle, the ability to reposition a gun under any condition could mean the difference between survival and destruction.

(**Opposite, above**) A 3.7cm *Pak* is positioned and ready to engage 'enemy' armour during training. The section commander scans the terrain ahead through his 6×30 binoculars, gauging distance and searching for the perfect moment to give the order to fire.

(**Opposite, below**) A 3.7cm *Pak mit Leicht Lafette fur Fallschirmjäger* with parachute positioned under the fuselage of a Ju-52/3m transport plane.

(**Above and below**) *Fallschirmjäger* (German parachute troops) with a 3.7cm *Pak* gun, photographed in February 1944 while operating in a ground combat role. Two notable uniform modifications reflect this shift in role: first, standard infantry hobnailed marching boots have replaced the paratroopers' original jump boots with moulded rubber soles and heels; second, the troops are equipped with metal gas mask containers, instead of the lightweight canvas versions originally issued to *Fallschirmjäger* to minimize injury during airborne operations. (Bundesarchive Bild 183-2000-0210-502)

Chapter Three

Anatomy of the 3.7cm Anti-Tank Gun

The 3.7cm *L/45* gun had a barrel length of 45 calibres, featuring sixteen right-hand increasing twist grooves with a pitch of 6°, and no muzzle brake. The designation *L/45* represented a method of weapon classification employed by the German military to describe a weapon's length in relation to its calibre (bore diameter). The *L/45* gun barrel was forty-five times the 37mm bore diameter, or 166.5cm (5.47ft) long.

The barrel consisted of a core tube, with a monobloc barrel introduced into production during the war, made from forged, hardened steel, rifled and jacketed. The bore was drilled and smooth in the rear part, forming the cartridge chamber, with a transition to the rifled section of the barrel. The absence of a muzzle brake resulted in a considerable recoil, ranging from 515–595mm. The barrel had a flange that connected it to the breech block. Initially, the flange was conical, later changed to a polygonal shape. The barrel's durability was between 4,500 and 5,000 shots.

The barrel was mounted on a slide, allowing movement during firing. The barrel could be easily detached and replaced. The breech block housing contained several components, including the breech block itself, an ejector (for shell casings) and a mechanism for operating the breech. The breech block was a horizontal wedge type. The barrel, along with its fittings (housing) and slide, was mounted in a cradle. The movement of the slide with the barrel in the cradle was guided by trunnions. The sides of the cradle had elements for mounting it in the upper carriage of the gun.

The cradle elements were connected to a worm gear elevation mechanism and a traverse mechanism. Two small hand wheels that operated the traverse and elevation of the gun were located on the left side of the barrel.

The cradle contained a recoil mechanism, consisting of a hydraulic recoil buffer and a spring recuperator. The cylindrical recoil buffer, with a capacity of approximately 1.1 litres, was filled with hydraulic fluid (most commonly Braun). The spring recuperator allowed the barrel and slide to return to their original position after firing, enabling the gun to be reloaded for the next shot.

The gun was semi-automatic (only for anti-tank shells), with manual loading of the cartridge into the chamber and firing using a button located on the left side of the breech. The automatic mechanism enabled the breech to open, eject the casing, close the breech and cock the firing pin, which was then released by pressing the button. The breech could be opened manually using the breech lever, which would also cock the firing pin.

In case of a misfire, there were three safety arrangements. Before the breech could be opened, the hand-operated block stop had to be released by pressing the plunger. When turned to the *Sicher* ('safe') position, the safety catch on the left prevented the sear from releasing the firing pin, and also prevented the breech from opening. The gun could not be fired when the buffer cylinder was not correctly assembled, as a spring-loaded plunger prevented the actuation of the sear.

The horizontal firing angle was 29.5° to each side, giving a total of 59°. Vertically, the firing angles ranged from -8° to +25°. When firing at enemy tanks, the maximum effective range was determined to be 800 metres. The *Carl Zeiss Jena Zielfernrohr ZF* telescopic sight was calibrated to these values.

The *Pak* had two triggers. One was a button, located inside the elevating wheel, worked by a trigger cable. The second trigger was a lever located on the lower right side of the breech block.

Sights: The *Zielfernrohr ZF* 1×11 sights instruments (a telescopic sight with 1× magnification and an 11° field of view) and the aiming mechanism were mounted on an upright bracket carried on the top of the carriage. Adjustment of the sights was obtained by two pairs of adjusting screws (one pair for vertical adjustment, the other for horizontal).

Trails: The gun had two detachable 7cm diameter aluminium alloy tubular trails (with the trails closed, the barrel traverse angle was 6°, with them open 60°), ending with spades and a towing hitch, used to stabilize the gun. These trails were attached to the lower carriage, which consisted of a base where the wheels were mounted and the upper carriage of the gun (connected to the lower carriage by trunnions) with the barrel, accessories, and elevation and traverse mechanisms.

The spade was 22.8×22.8cm when in the closed position, and had a hardened steel tip to assist the weapon dig into the ground when fired.

A handle was located above each side of the spade to assist the crew in moving the weapon into position, or for portage. A tow ring was mounted on the end of the right leg of the trail.

Shield: The armoured shield was made from surface-hardened (by carburizing) armour plate, with a thickness of 4.5–5mm. It consisted of two central elements and four side elements, which were mounted with brackets to the upper carriage. The individual elements were connected by hinges, allowing for the sides of the gun shield or the three upper shield elements to be folded, thereby reducing its height if needed. When the shield parts were positioned, the elements were locked with special latches. There was also the option to completely remove the armoured shield.

The leg shield folded under the lower carriage when travelling and folded down to the ground level when in action. The gun shield had a slope of approximately 45° when in action, the upper half being hinged and folded forward as required.

Wheels: Independently suspended wheels had low-pressure Continental tyres featuring an off-road tread. The wheel discs were solid, stamped from 3mm thick sheet metal, and attached to the hub with five bolts. The wheels had spring shock absorbers. Guns equipped with pneumatic tyres were designed for towing by mechanical traction at speeds of up to 40–45 km/h.

The gun was equipped with a barrel cover, a muzzle cap used during transport and a breech cover. A collapsible rod was used for cleaning purposes.

Specification

Designation	**3.7cm *Pak 35/36 L/45***
Originator and Producer	Rheinmetall-Borsig AG, Dusseldorf
Years (*Tak/Pak*)	1930–1941
Calibre	37mm (1.46in)
Length of tube	1.665m (65.5in)
Overall width	1.65m (65in) (1.85 *L/50*)
Overall height	1.22m (48in)
Overall length	3.44m
Barrel length	1,664.8mm (1,850 *L/50*]
Gun barrel weight	97kg

Rifling length	1,308mm
Theoretical rate of fire	16–20 shots/min
Rate of fire	12–15 shots/min
Transport in action	444kg (979lb)
Gun weight on firing position	328–337kg
Carriage	split trail
Wheel track	1.54m (60.6in)
Wheel type/diameter	steel with pneumatic tyres
Breech type	sliding wedge
Recuperator type	hydro-spring
Elevation	-8° to +25°
Horizontal	*Tak* 20°+20° / *Pak* 30°+29°
Fire height	610mm
Maximum recoil	595mm
Traverse	59° trails open; 4° 30' trails closed
Muzzle velocity	745m/s (2,444fps)
Maximum range	6,800 metres (7,437 yards)
Effective range	800 metres (875 yards)
Rate of fire	12–15 rounds per minute

Production

The first thirteen pre-production units of the *Tak 28* were manufactured by May 1929, and serial production of the guns began in 1930. By 20 November 1932, a total of 264 3.7cm anti-tank guns had been delivered, with 201 in the *Reichswehr* and sixty-three in the supplementary inventory.

The Treaty of Versailles limited the *Reichswehr* to just seven infantry divisions and three cavalry divisions (totalling 100,000 soldiers), and the overall number of new anti-tank guns needed for such a force was limited to a few hundred. However, in 1934, the *Reichswehr* began a secret expansion programme to 300,000 men in twenty-one divisions, increasing the need for a large number of anti-tank guns and boosting production.

Mass production of *Tak/Pak* guns in Düsseldorf started in 1936, following the merger of Rheinmetall AG and August Borsig GmbH under the name Rheinmetall-Borsig AG. By the start of the war in September 1939, the factory had produced approximately 11,200 *Pak* guns, and by the end of production of this model in 1942, around 16,500 units had been made.

According to records in the 'Overview of Ground Forces Armament' (*Überblick über den Rüstungsstand beim Heer*), following the beginning of the Second World War, *Pak* guns were produced in the following amounts:

Month/Year	1939	1940	1941	1942
January	?	251	122	4
February	?	271	192	13
March	?	265	178	15
April	?	277	195	–
May	?	274	109	–
June	?	246	183	–
July	?	210	114	–
August	?	228	235	–
September	300	207	95	–
October	271	150	39	–
November	287	192	3	–
December	371	142	0	–

The record from 1 March 1942 states 'order completed'. Overall, 16,639 guns were built during the period of mass production.

Ammunition

The 3.7cm *Pak* guns fired armour-piercing (AP) and high-explosive (HE) fragmentation shells. According to German nomenclature, the name of a cartridge consisted of the name of the projectile and the added word '*Patrone*', for example, *Panzergranate Patrone 40* (*Pz.Gr. Patr. 40* or *Pz.Gr. 40*) or *Sprenggranate Patrone 40* (*Sp.Gr. Patr. 40* or *Sp.Gr. 40*).

***Panzergranate* (*Pz.Gr. 36*):** full-calibre armour-piercing (APHE) shell with a length of 119.2mm and projectile weight of 685g. The shell was uncapped and was fitted with a single copper driving band. Inside, there was 13g of explosive charge, consisting of pentrite (PENT) with a 15 per cent wax addition. Pentrite in its pure form is very sensitive to external stimulation; the wax was added to act as a desensitizer, reducing this sensitivity. The base of the shell was fitted with a *Bd. Z. 5103* (or *Bd. Z. 5103**) fuse integrated with a *Leuchtspurhülse Nr 1* tracer, which had a mass of 23g, and its burning time was two seconds. There was also an initiator, *Sprengkapsel fur 3.7 der Pz.Gr.*, weighing 1g. The initial velocity of this projectile was 745m/s. It could penetrate armour (angled at 30° from the vertical) of 35mm thickness at 100m and 28mm thickness at 500m. Theoretically, it could penetrate 22mm-thick armour at 1,000m, but such distances were rarely targeted.

In 1939, an improved projectile with a stronger bursting charge (the rounds were compound) was introduced, the *Pz.Gr. 39*. This weighed 0.69kg and achieved an initial velocity of 760m/s. At a distance of 100m, it could penetrate armour 45mm thick, and 36mm thick at 500m.

Armour penetration table for 3.7cm *Pak/Kwk*

	Range					
	100m	**200m**	**400m**	**500m**	**1,000m**	**1,500m**
Pz.Gr.	34mm			29mm	23mm	19mm
Pz.Gr. 40	64mm	53mm	38mm	31mm	–	–
Stielgr. 18	34mm			28mm	22mm	19mm
Stielgr. 41	180mm	180 mm	–	–	–	–

***Panzergranate 40* (*Pz.Gr. 40*):** first model of sub-calibre projectile (APCR) with a tungsten core (*Hartkerngeschoss*), which had a spool shape with two guiding protrusions; its mass was 368g and it achieved an initial velocity of 1,020m/s. Thus, at a distance of 100m, it could penetrate armour 64mm thick, and at 500m still 31mm thick, but its effectiveness decreased at greater distances. It was 1.5 times more efficient in armour-piercing capabilities than the previous *Pz.Gr. 39*.

A ballistic cap made of either plastic or an alloy of aluminium, copper and manganese was mounted on the body. In the first case, the total length of the shell was 305.5mm, while in the second it was 88.3mm. The caps differed slightly in shape, the metal one being more slender but shorter. The space between the cap and the core was filled with plastic. The tungsten carbide penetrator was 58mm long and had a diameter of 16mm. The shell did not contain any explosive material or fuse. A tracer was placed at the bottom. Even after penetrating armour, the shell was not always able to effectively damage the interior of the hit vehicle.

In encounters with well-armoured vehicles (French in 1940 or Russian in 1941), the gun crews of the 3.7 cm *Pak* were left to aim at the opponent's vulnerable points (hatches, optics, running gear, side air intakes etc.), hoping to cause at least some damage.

Sprenggranate 18 umgearbeitet (Sp.Gr. 18umg.): high-explosive (HE) fragmentation shell with a length of 134mm and a weight of 685g (including the *A. Z. 39* fuse), and an initial velocity of 745m/s. The shell comprised the aluminium container, which in the upper cavity contained a pressed pellet of flake pentrite (26g) mixed with wax, in a ratio of 90–10, while the lower cavity contained a comparatively large tracer. The weight of explosive material was comparable to that of a hand grenade. The shell body was encircled by a single driving band made of copper or soft iron.

The propellant was in tubular form, consisting of smooth black tubular sticks. The composition of the propellant consisted of: nitro-cellulose 65.12 per cent (Nitrogen content 12.14 per cent), diethylene glycoldinitrate 31.52 per cent, centralite 2.65 per cent, potassium sulphate 0.42 per cent and graphite 0.29 per cent.

Ammunition types and sub-types

Type of ammo	***Length of shell/ projectile (mm)***	***Weight of shell/ projectile (kg)***	***Fuse***	***Length of shell/ projectile (mm)***	***Weight of explosives (g)***	***Weight and type of propellant (g)***	***Muzzle velocity (m/sec)***
***Pz.Gr.* [HE-T*]**	337.0/119.2	1.32/ 0.685	*B.d.Z.5103**	337.0/119.2	13	Digl.R.P. 189	760–820
***Pz.Gr. 40* [APCR-T]**	305.5/88.3	0.97/ 0.368		305.5/ 88.3	–	Digl. Step. 186 /RJP.150	1,020
***Sp.Gr. 18* [HE-T]**	341.5/134.4	1.26/ 0.615	*A.Z. 39 3.7 cm KpfZ Zerl.P*	341.5/ 134.4	26	Digl. R.P. 175	745
***Sp.Gr. 40* [HE-T]**	353.5/140.0	1.31/ 0.665		353.5/ 140.0	45	Digl. R.P. 175	745
***Stielgr. 41* (HEAT)**	738	8.51	*A.Z. 5075 + B.d.Z. 5130*	738	2,420	Ngl. R.P. 217	110
Pz.Gr. (Üb)	?	0.97		?	–	Ngl. R.P. 140	?

The igniter consisted of a circular silk bag attached to the base of the charge by a silk tape and contained 34.4 grains of the usual greyish-green porous nitrocellulose powder in the form of short cylindrical grains. The composition consisted of: nitrocellulose (including graphite) 87 per cent, diethylene glycodinitrate 11 per cent, centralite 0.5 per cent, diphenylamine 0.5 per cent and 1 per cent of volatile matter.

Four types of fuses were used interchangeably: 3.7cm *Kpf. Zerl. P.*; 3.7cm *Kpf. Zerl.Py.*; *A.Z. 39*; *A.Z. 39 Zn.* Additionally, there was also a tracer, *Leuchtspurhülse Nr 3*.

The shell was equipped with the primer C/13 n.A.. The amount of mercury fulminate in the cap composition had been decreased and granular gunpowder had been introduced between the cap and the pellet of pressed gunpowder. The cap composition consisted of: mercury fulminate 30.6 per cent, potassium chlorate 35.2 per cent, antimony sulphide 27 per cent and ground glass 7.2 per cent.

Sprenggranate 40 (Sp.Gr. 40): high-explosive (HE) fragmentation shell with a single driving band made of copper or soft iron. Compared to their predecessors, *Sp.Gr. 18umg*, the length increased to 140mm and the weight to 665g. A significantly smaller tracer, *Leuchtspurhülse Nr 1*, was used. This allowed for an increase in the volume of the explosive charge chamber, which enabled the inclusion of 45g of pentrite mixed with wax (in a ratio of 90–10). This was 19g more than in the case of the *Sp.Gr. 18umg*. Two types of fuses were used: *A. Z. 39* and *A. Z. 39 Zn*. The introduction of the *Sp.Gr. 40* improved the effectiveness of fire, although it still wasn't very high (which is understandable given the 3.7cm calibre). The HE ammunition was used to combat small targets, such as machine-gun nests or individual guns, as well as to eliminate unarmoured vehicles or attempt to damage armoured vehicles.

Stielgranate 41 (Stielgr. 41): separate-loading ammunition (HEAT), meaning the projectile and the case with the propellant were not a single unit. Interestingly, they were loaded into different ends of the barrel. The *Stielgr. 41* somewhat resembled mortar grenades in shape. These were over-calibre projectiles (projectile calibre larger than barrel calibre) with a length of 740mm and a weight of 8.51kg. The warhead contained a powerful shaped charge, equipped with a characteristic hollow.

The explosive material consisted of 2.4kg of RDX with TNT (in proportions 50–40). To reduce the likelihood of a dud, two fuses were installed: the *A.Z. 5075* (in the nose of the projectile) and the *Bd.Z. 5130* (in the base of the charge). There were also two boosters. A long rod extended from the warhead (which was inserted into the barrel from the muzzle), along with a perforated sleeve (encompassing the barrel from the outside). To ensure stabilization, the projectile was equipped with six stabilizing fins. The advantage of this design was the lack of rotational movement during flight, thus avoiding the dispersion of the shaped charge jet. The *Stielgr. 41* was fired by a propellant charge placed in a standard cartridge case sealed with two protective discs. When loaded, the cartridge case was in the chamber, while the projectile's warhead protruded from the barrel's muzzle.

Ammunition production quantity, 3.7 cm *Pak L/45*

Type of ammo	1939	1940	1941	1942	1943	1944–45	Total
Pz.Gr.	1,300,000	5,623,000	3,200,500	2,560,000	1,250,000	–	13,933,500
Pz.Gr. 40	–	286,600	885,200	207,200	1,000	–	1,380,000
Sp.Gr. 18 umg	400,000	4,336,300	1,059,300	2,526,700	2,893,500	–	11,215,800
Stielgr. 41	–	–	–	600,900	35,100	–	636,000

Despite the *Stielgr. 41*'s impressive destructiveness and significant effect inside a hit vehicle, they were not effective in combat. The low muzzle velocity resulted in poor accuracy, and the effective range was up to 300m at best. To make matters worse, loading the projectile into the barrel required the soldier to expose himself from behind the armoured shield, significantly exposing him to fire during combat. Destroying an enemy vehicle was only possible under favourable circumstances, such as from an ambush, where there was a good chance of hitting with the first shot from close range. In this type of projectile, malfunctions of propellant charges often occurred, further reducing the effectiveness of the *Stielgr. 41*.

For training purposes, 3.7cm *Pz.Gr. Üb* cartridges without projectiles were used. During training, dummy ammunition was used because of the lack of gas pressure in the barrel, which prevented the gun's automatic mechanism from functioning. Therefore, muzzle plugs were used to enable the operation of automatic components.

Ammunition was transported in metal boxes or containers holding twelve rounds, weighing between 20.5kg and 20.9kg (depending on the type of projectile and whether wooden dividers were installed inside the box). The towed single-axle ammunition trailer, *Sd.Anh.32*, typically carried 180 anti-tank and fragmentation shells.

Gun crews underwent training using specialized carbine barrels inserted into the gun barrels, which allowed them to fire 5.6mm sporting ammunition. This simulation method was crucial for familiarizing gunners with the handling and operation of the weapon system in a controlled and safe environment before actual live-fire exercises or combat situations.

Variants

3.7cm as a tank gun – *Kampfwagenkanone* (Kwk)

The Treaty of Versailles restricted Germany from developing and possessing tanks. However, due to a secret agreement with the Soviet Union, resulting from the Treaty of Rapallo, the *Reichswehr* established a tank testing facility named the *Panzertruppenschule Kama*, located near Kazan in the Soviet Union, in 1926. This location served as a testing ground and tank training facility for both the Red Army and the *Reichswehr*, where German tank prototypes, disguised under the name 'tractors', were secretly tested. The 3.7cm *Pak* gun was a natural choice for arming one of those early armoured fighting vehicles conceived as a tank destroyer.

The general requirements for the development of this tank prototype – codenamed *Leichttraktor* (light tractor) – were outlined in the summer of 1928. Specifications included a 360° rotating turret mounting a (semi-) automatic 3.7cm gun with a minimum ammunition capacity of 150 rounds, accompanied by a coaxial heavy machine gun.

After the four prototype vehicles were completed in April and May 1930, they were transported to the Kama training ground. There, they underwent extensive testing that included mobility and mechanical endurance trials, and also comprehensive gunnery exercises with the 3.7cm *Pak* gun. The prototypes remained at the facility for over three years, only returning to Germany in the autumn of 1933.

They were then turned over to the Army Equipment Office in Spandau, which sent them on to the Panzer Gunnery School of Alt-Gaarz at Wustrow at the end of the year. Without any major treatment of their weak points, they were kept ready for tactical and firing training. In this role they remained extremely useful for a number of years.

While the *Leichttraktor* project was of limited success, the tests at the Kama centre resulted in enthusiasm for the building of a multi-turreted tank called the *Neubaufahrzeug* (new-built vehicle). A contract was issued in October 1932 to build a 15/20-ton tank with the main turret armed with a 7.5cm tank gun and a 3.7cm support gun with just 37 rounds. In total, only five completed vehicles of this type were delivered by early 1936.

The *Neubaufahrzeug* was hard to manage because of the large crew and the problems of coordinating the multiple turrets. It also had an underpowered and unsuitable powerplant and running gear that no longer represented the technical potential of the times. Due to those issues, none of the *Neubaufahrzeug* were assigned to front-line units but used for training at the Putlos Gunnery School. Interestingly, three of those machines saw action in April 1940 in Norway.

Despite the disappointments resulting from the development of the tank-tractors, the 3.7cm gun remained the only available choice for the development of much-needed *Zugführerwagen* (*ZW* – a platoon leader's vehicle, later known as the *PzKpfw. III*) for the newly formed panzer divisions in the mid-1930s.

At the end of 1934, Daimler Benz was contracted to develop a test chassis for this new vehicle, but the exact configuration of the suspension was still undecided. The *ZW* was proposed to be a 10–12-ton class tank, but the actual production version weighed 15 tons. Krupp fabricated the turret, which mounted a 3.7cm *Kwk L/46.5* and two *MG 34* machine guns, that equipped *PzKpfw. III Ausf. A* to *G* models.

At this stage it was believed that the 3.7cm *Pak/Kwk* could fight any modern tank. However, already in 1936, development of 5cm and 7.5cm anti-tank guns was on the way. Interestingly, in the initial design stage, the turret ring of the *PzKpfw. III* was made large enough to accommodate a 5cm gun should a future upgrade be required.

When the prototype *PzKpfw. III*s were made available to the *Wehrmacht* in 1937, due to their shortcomings and deficiencies of trial series, the mass production had to be delayed until 1938.

Production of *PzKpfw. III* equipped with 3.7cm *Kwk 36 L/46.5* AT gun

***Ausführung*/Model**	**A**	**B**	**C**	**D***	**E**	**Total**
Year	1937	1937	1937–38	1938–40	1939	
Produced	10	10	15	25+5	96	161

* Note that the D1 model was equipped only with a 3.7cm wooden mock-up gun

A total of eighty-seven *PzKpfw. III*s of all types saw combat use in the invasion of Poland in September 1939. These were deployed primarily with *1. Pz.Div.* (1st Panzer Division), with twenty-six tanks, and *3. Pz.Div.* (forty-three tanks), while the remaining divisions had only a token number or none at all. Combat losses were relatively high, with total losses of twenty-six *PzKpfw. III*s (30 per cent), yet in the opinion of *Panzertruppen* their '3.7cm guns had great accuracy and proved to be effective against enemy tanks'.

The after-action reports, however, requested a high-explosive round for the 3.7cm *Kwk* be used against soft targets. Another recommendation was to replace the 3.7cm by a 5cm or 7.5cm anti-tank gun. Indeed, starting from the *PzKpfw. III Ausf. F* model, the 5cm *Kwk 38 L/42* gun was introduced.

3.7cm as a self-propelled gun

The first tests of equipping the *Tak* gun with a tracked chassis took place as early as 1929. During these tests, the 3.7cm gun, without its armoured shield, was mounted on the chassis of a WD-type half-track tractor. Apparently, these tests were not successful, as the self-propelled gun in this configuration did not go into serial production.

During the French campaign (May–June 1940), some anti-tank companies created improvised anti-tank guns by mounting the 3.7cm *Pak* gun on the body of a *Kfz. 69* wheeled artillery tractor without removing the wheels or making any other modifications.

In June 1940, the 44th Mountain Anti-Tank Battalion (*Gebirgs Panzerjäger Abt. 44*), part of the 1st Mountain Division (*1. Gebirgs Div.*), had a *Kfz. 69* tractor that was modified by mounting a special frame with a rotating base for the *Pak* gun. Two other Krupp tractors in the same unit were also armed with 3.7cm guns. However, in their case, the driver's cabin and engine hood were protected by 8mm-thick armour plates.

Over time, the *Pak* was mounted on several types of mass-produced armoured transport carriers to increase its mobility and effectiveness. The most well-known variant featuring the *Pak* gun mounted on an armoured half-track was the *Sd.Kfz. 251/10*, which appeared as early as the French campaign in 1940. One such support vehicle was allocated to each platoon commander (*Führerzug* in a *Panzer-Grenadier Kompanie*) in units equipped with medium armoured personnel carriers (*Mittlerer Schützenpanzerwagen*), to fire on enemy armoured fighting vehicles or hard targets. Initially an ingenious tinkering in some battalions, it became an indispensable part of the *Sd.Kfz. 251* platoons up to 1944.

The first models were obtained by fitting the complete *Pak* cradle (with its full shield) onto the armoured transport carriers, using an added axle on the crew compartment roof, without the wheel axle and trails. The most noteworthy modifications were made to the shield. Initially, it was the regular shield, but its large dimensions made the vehicle too conspicuous and hindered the gun commander's vision. Consequently, the shield was cut down on the *Sd.Kfz. 251 Ausf. A–C*. Later, only a half-shield was installed on the left to protect the gunner on the *Ausf. B–Ds*. On the last models (*Ausf. C–D*), the mounting layout was modified to lower the silhouette of the gun further, with the shield either removed or offering only partial protection on the sides.

The *Sd.Kfz. 251/10* gained fame on the Eastern Front as a 'sniper killer', with the precise aiming mechanism and flat trajectory of the 3.7cm *Pak* shells making it highly effective in direct fire support. A telescopic *Zielfernrohr* 2x16° field of view sight was used to aim the gun.

The vehicle was crewed by five or six men and carried 168 shells for the gun. It was also equipped with a machine gun and a *Pz.B.39* anti-tank rifle.

The production of the *Sd.Kfz. 251/10* was only halted in November 1943, with the last 231 units rolling off the production lines that year. In 1944, the remaining *Sd.Kfz. 251/10*s still in service were withdrawn from front-line units, and some were likely rearmed with 7.5cm *Kwk 37* guns.

Pak guns were also mounted on light half-track armoured transporters, the *Sd.Kfz. 250/10* model, which were mass-produced for the *Wehrmacht* from mid-1941. The production of this variant of the transporters was halted in November 1943, with the last 151 units produced that year.

The Germans also mounted the *Pak* on various captured tracked vehicles to create mobile anti-tank platforms for infantry divisions. These improvised solutions were mostly used on the Eastern Front, where the need for off-road mobile anti-tank units was particularly acute.

Among these vehicles, the most numerous was the *Selbstfahrlafette für 3.7cm Pak*, based on captured French Renault UE light artillery tractors (*chenillette* in French).

There were occasional instances of using captured *chenillettes* to create improvised self-propelled anti-tank guns during the 1940 campaign in France. Archival photos show that the *Pak* gun was mounted on the hull of the tractor without removing the wheels or making any other modifications. The wheels of the gun were placed in the folds of the front fenders of the Renault UE, and the trails of the lower carriage were attached to the cargo box. It is hard to imagine a simpler solution. Additionally, to avoid being mistakenly fired upon by their own troops, large national markings (*Balkenkreuz*) were painted not only on the hull of the tractor but also on the protective shield of the gun.

Out of 2,500 UE and UE2s captured by the *Wehrmacht*, approximately 700 were converted into *Pak* carriers by mid-1941 after an overhaul by the Ateliers de Construction d'Issy-les-Moulineaux (AMX) factory under the guidance of the German MAN company. This time, the tractors were mounted with *Pak* guns featuring the upper carriage and original protective shield, but without removing the lower carriage. There was no room for the crew in such a small vehicle, so the gun had to be operated while standing behind it. This new vehicle was designated as the 3.7cm *Pak (Sf) auf Infanterieschlepper UE (f)*.

The Renault UE could also function in anti-tank units in their original primary role as munitions carriers, known as *Munitionsschlepper Renault UE(f)*. Some of these had an armoured roof fitted above the bin to protect the ammunition load against overhead shell airbursts.

The new half-French/half-German self-propelled guns were mainly assigned to the 11th and 12th Wave (*Welle*) infantry divisions, in which they fought on the Eastern Front from the summer of 1941. It is worth noting that, according to some studies, the 3.7cm *Pak (Sf) auf Infanterieschlepper UE (f)* were also assigned to independent tank destroyer battalions *(Panzerjäger Abteilungen)* with the following numbers: 463, 525, 560, 563, 652 and 654. Each of these battalions theoretically should have had thirty-six vehicles, totalling 216 self-propelled guns.

Even some ex-French light infantry tanks, like the R-35, were also modified to carry the 3.7cm *Pak* and pressed into German service as either second-line tanks or support vehicles.

The *Pak* guns were also mounted on British tracked carriers captured during the Battle of France, including the Bren Gun Carrier, Universal Carrier and Scout Carrier, designed as 3.7cm *Pak 35/36 auf Fahrgestell Bren(e)*, 3.7cm *Pak auf Selbstfahrlafette Bren Carrier (e) 731 and* 3.7cm *Pak Fahrgestell Bren(e)*. Vehicles of this type probably equipped training units *(Panzerjäger-Ersatz-Abteilungen)* stationed in Germany

Lastly, after the initial success of the German invasion of the Soviet Union, resulting in the capture of a large number of *T-20 Komsomolets* light truck gun carriers, these machines were also converted into mobile *Pak* platforms designated as *Pak auf Gepanzerter Artillerie Schlepper 630(r)*.

All of these conversion efforts reflect the improvisational nature of German mobile artillery during the early stages of the war, when more purpose-built tank destroyers were still in development or in short supply.

Artillery Tractors

Although the primary focus of this work is on the 3.7cm gun itself, it is pertinent to discuss the towing vehicles used with *Paks*. Initially, the *Tak* guns were towed by horses, but this method was quickly replaced by motorized traction. This transition was made possible by the *Reichswehr*'s motorization programme announced in 1926, under which renowned companies were contracted to develop cross-country staff cars and light cargo lorries capable of towing light guns.

Initially, the programme favoured six-wheeled cars with partial all-wheel drive for troop transport. However, this concept was abandoned during trials due to high production costs and the limited capacity of the German automobile industry at that time. Consequently, the *Reichswehr* explored the use of existing civilian models with rear-wheel drive, solid axles and four-stroke gasoline engines. Minor modifications to the chassis provided these vehicles with the necessary off-road capabilities. To accommodate quick troop mounting and dismounting, a specialized superstructure was required.

Under the guidance of the well-regarded chassis manufacturer Trutz from Coburg, the distinctive *Kübelsitz-Wagen* (bucket-seat field car) was developed, affectionately nicknamed the *Kübelwagen* by soldiers. Introduced in 1927–28, early models were produced by Horch, Adler and Mercedes-Benz, with additional manufacturers such as BMW, Hanomag, Phanomen, Stoewer and Wanderer joining the production effort soon after. Various sub-contractors or automobile plants themselves built the vehicle bodies. Most of those types were used as *Pak* tractors already in the early thirties.

As part of the initial motorization programme, contracts were also awarded for the production of off-road-capable three-axle (6×4) trucks. In response, the *L2H43/143* truck debuted in 1936. This versatile vehicle served as both a light truck and a motorized limber for the 3.7cm *Pak* guns. By the mid-1930s, the Krupp L2H43 emerged as the most common towing vehicle for light guns and was widely used under the designation of *Krupp Protze Kfz. 69.*

Also critical to the transportation of 3.7cm guns was the single-axle ammunition trailer type *SdAh. 32*, towed by *Kübelwagens* or *Krupp Protze* tractors. These trailers, essential for transporting ammunition, featured a sheet steel box capable of holding sixty-four rounds without cases in an upright position. The *SdAh.32/1* trailer was not armoured and cost only 668 Reichsmarks (RM).

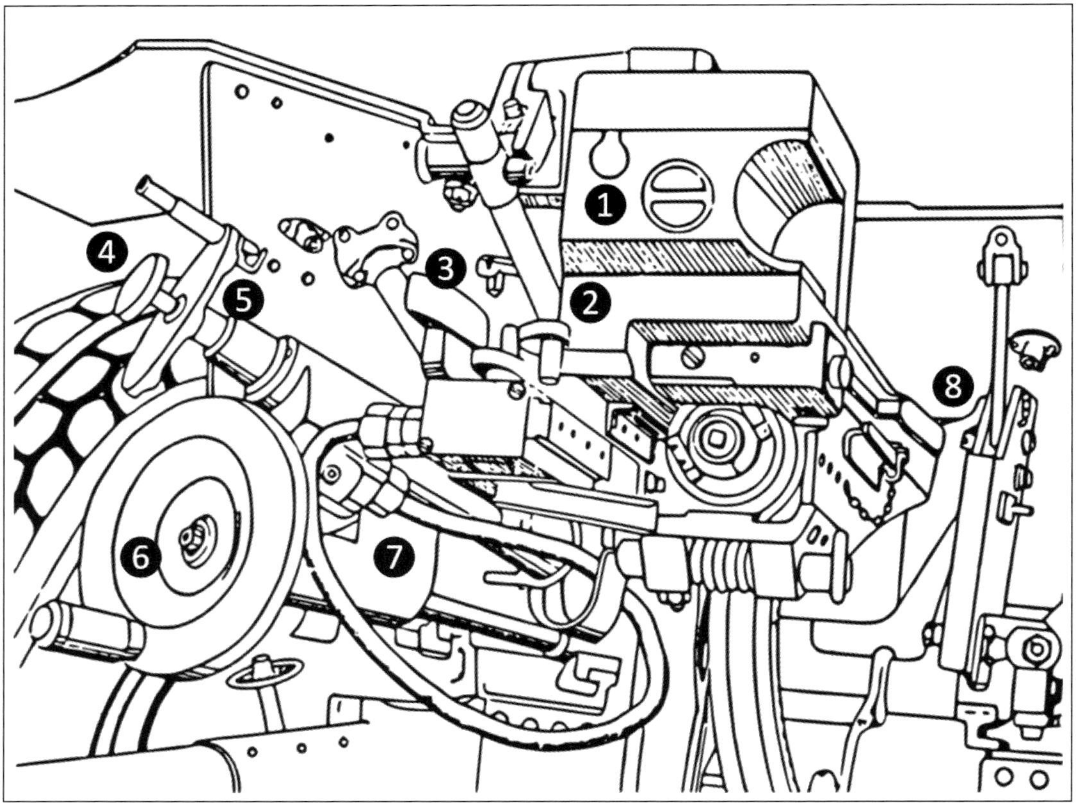

(**Above**) (1) Hand-operated block-stop plunger, (2) Trip lever, (3) Recoil recocking guide, (4) Left trigger, (5) Elevating handwheel, (6) Traversing handwheel, (7) Trigger cable, (8) Right trigger.

(**Opposite, above**) Soldiers of a *Landesschützen Bataillon*, based in Sosnowiec – a Polish city annexed into the German Reich in November 1939 – display their equipment in public.

(**Opposite, below**) Maintenance work typically included checking the breech mechanism. The horizontal sliding breech block had to be inspected for smooth movement, cracks or dirt. Lubrication was applied sparingly to prevent seizing.

The small hydro-spring recoil system also needed regular checks for leaks and sufficient fluid/oil pressure. Springs were inspected for wear or deformation.

(**Above and below**) The firing pin was particularly prone to wear, carbon buildup and fouling from repeated use. Its routine servicing ensured the reliability of the *Pak*, especially during prolonged exercises or combat operations.

A Pak section approaches its designated firing position during pre-war manoeuvres. The towing vehicle, most likely a Mercedes-Benz 260 *Kübelwagen*, carries temporary tactical symbols: removable plates mounted on the left rear mudguard and above the front bumper. Vertical stripes marked the 3rd Company, a system designed for clarity in fast-moving exercises.

Pak armed with the *Stielgranate 41*. Its range was short and accuracy limited, so it was typically used in ambush positions, hidden in hedgerows or tree lines where tanks would approach within striking distance.

(**Above**) Gun crews conduct post-combat cleaning in France, probably within the grounds of a large farmstead pressed into service as a temporary staging area. Such buildings were commonly repurposed by German troops during the campaign, before they later moved into the barracks vacated by defeated Polish forces.

(**Opposite**) The upper part of the protective shield was designed to be folded down during field manoeuvres, allowing the crew to maintain a clear, unobstructed line of sight. This improved their ability to accurately observe enemy movements at longer ranges. This feature was particularly useful during training sessions, where precision and effective targeting were crucial for developing the gunners' skills.

The red elastic bands seen on helmets were used during field exercises for 'friend or foe' identification, helping to prevent confusion and ensure safety as manoeuvres were carried out.

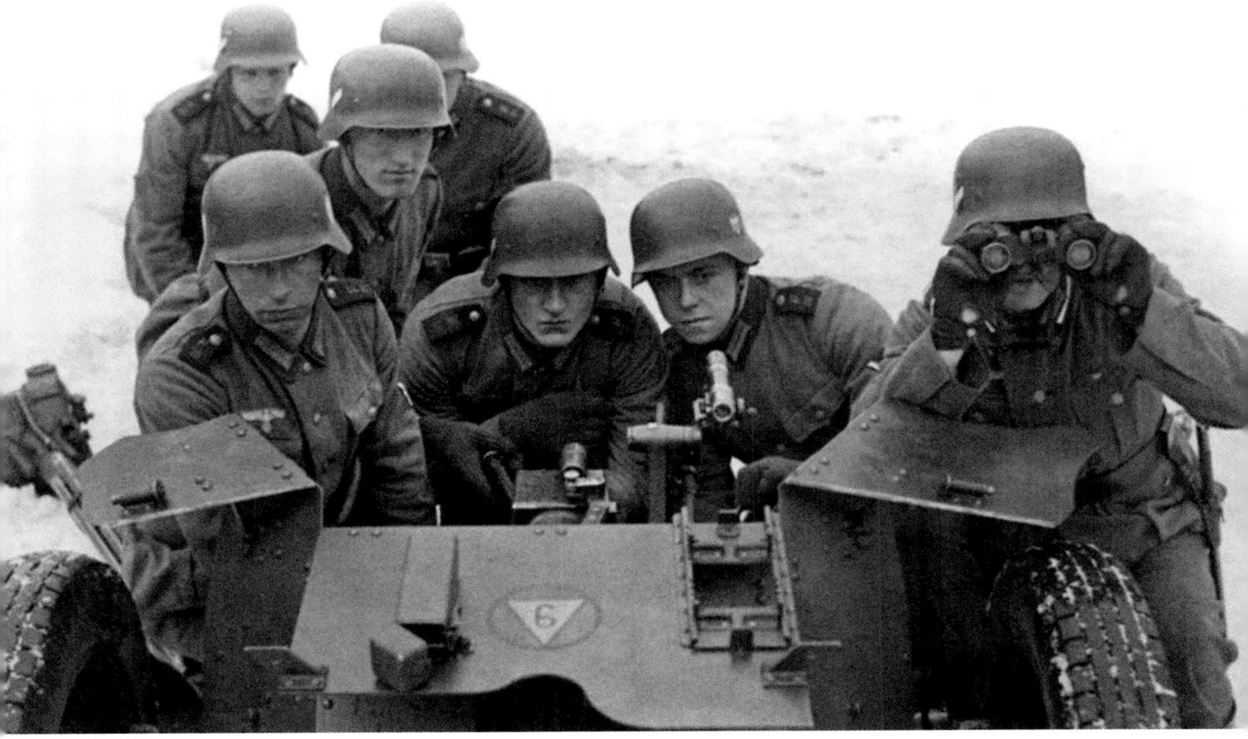

A 3.7cm *Pak*, viewed from all sides.

A *Pak* photographed in the summer of 1941 in Russia. The gun commander had to use binoculars to detect targets from a long distance, the gunsight being unable to aim beyond 800 metres.

A *Pz.Abw.Abt.* is ready for inspection in mid-1937. The unit was equipped with Adler Type 3 Gd cross-country cars, powered by a six-cylinder, 60-horsepower engine, used as gun and ammunition trailer tractors. Nearly 4,300 units of the Type 3 Gd were produced between 1936 and 1940.

A *Pz.Abw.Abt.* during a street parade in Günzburg, Bavaria, in the autumn of 1937. The anti-tank battalion was awaiting the arrival of the *Krupp Protze* six-wheel vehicles, which were to be used as dedicated gun tractors. However, due to delays in delivery, it had to rely on vehicles like the Opel Blitz 1.5t.

A *Pz.Abw.Abt.* on manoeuvres, likely conducted at the training grounds in Neuruppin in 1938. Note the wide tow belts worn across the soldiers' chests, used to move guns into firing positions. Interestingly, while all the soldiers are wearing traditional infantry helmets, the one on the left has the so-called cavalry helmet (M.1918), which features ear cutouts.

The 3.7cm *L/45* gun had a barrel length of 45 calibres, featuring sixteen right-hand increasing twist grooves with a pitch of 6°, and no muzzle brake.

A view of the *Pak* production line at Rheinmetall-Borsig AG in Düsseldorf, showing the assembly and manufacturing process.

A close-up of a fully assembled late-version *Pak* gun, highlighting its design details; note that the gunsight has not yet been installed.

A tank version of the *Pak* gun, called *Kwk*, mounted centrally with the *Turmzielfernrohr* (turret sighting telescope) on the left-hand side of the breech block of a *Panzer III*. On the right side are two coaxial 7.92mm *MG 34* (Model 1934) machine guns.

During the French and Soviet campaigns, anti-tank teams were not always provided with adequate support from ammunition sections. As a result, they often had to carry additional ammunition themselves, frequently attaching ammo boxes directly to the gun's splinter shield.

An anti-tank section of the *290. Inf.Div.*, photographed on the Eastern Front in the summer of 1941.

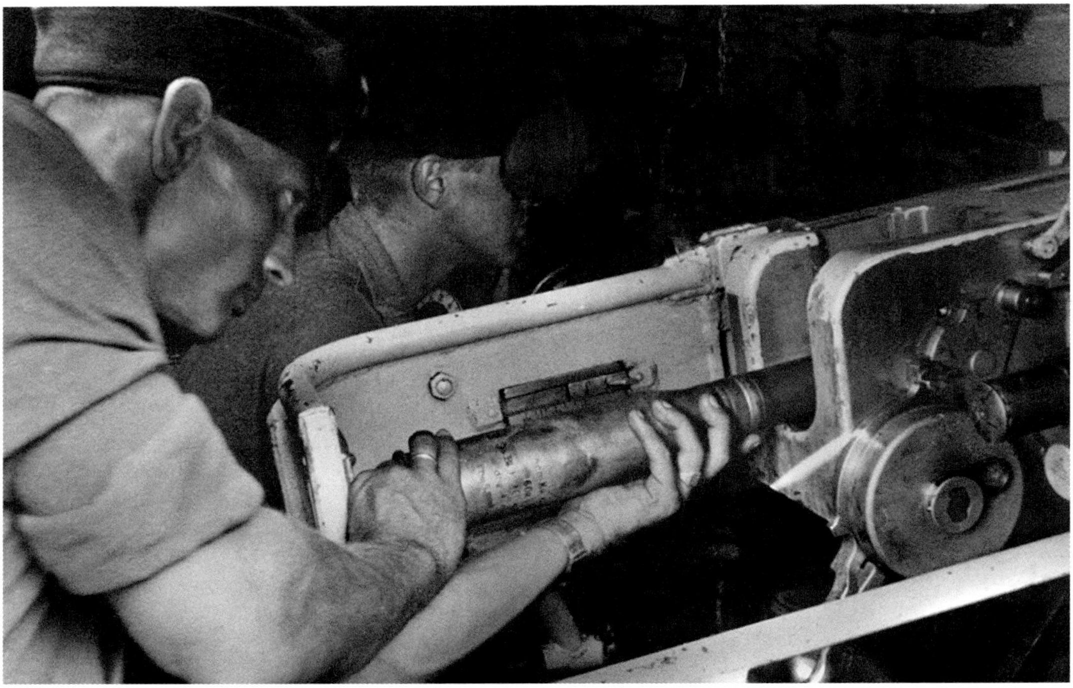

The loader of a *PzKpfw. III* is about to insert a round in the gun breech of a 3.7cm *KwK L/45*. The gun had a barrel length of 45 calibre and weighed 195kg. Elevation range was from -10° to +20°. Ammunition capacity varied by variant: 150 rounds (*III Ausf. A*), 121 rounds (*III Ausf. B–D*) and 131 rounds (*III Ausf. E–G*).

Wehrmacht officers inspecting the gun drill of a 3.7cm *Pak*. Note the ammunition boxes slung over the gun shield with clearly visible ammo designation (*Pz.Gr = Panzergranate*).

A 3.7cm *Pak* in a firing position. Compared to larger-calibre anti-tank guns, the 3.7cm rounds were lightweight and compact, which allowed crews to carry more ammunition and maintain a higher rate of fire.

A *Pak* crew loading their gun during a combat mission. One of the men handling ammunition can be seen wearing a Red Cross armband. His presence here likely indicates that the crew was understrength and had been temporarily assigned to fill the gap. Such ad-hoc substitutions were not uncommon during combat missions, when shortages of trained crew members had to be covered in order to keep the battery operational.

(**Above**) Members of an anti-tank platoon receive newly issued guns within a freshly prepared staging area. Similar wooden structures were erected to serve as parking bays for the accompanying *Kübelwagens*. The brick-and-mortar garages were constructed to serve as workshops rather than storage facilities.

(**Opposite, above**) A Krupp *Kfz. 69* artillery tractor, seen during field exercises, operates with its windshield folded flat and covered – standard practice to shield the glass from stones and debris kicked up by the wheels. Anti-tank regulations recommended this procedure whenever driving off-road. With no roof covering the crew compartment, soldiers endured dust, heat and glare, relying on their protective goggles to safeguard their eyes under the harsh summer sun.

(**Opposite, below**) A *Pak* gun crew positioned close to their weapon during winter peacetime exercises, practicing readiness and coordination under harsh conditions.

(**Above**) A close-up view inside *Pak* ammunition boxes reveals a well-organized storage system designed to hold a total of ten 3.7cm calibre shells each. The shape of the projectiles is flat and cylindrical, which was a characteristic feature of *Sp.Gr.* or high-explosive fragmentation shells.

(**Opposite, above**) This photograph captures a scene from the hot summer of 1941 in Lithuania, showing a *Pak* crew of the *121. Inf.Div.* in bivouac during early stage of Operation Barbarossa. A soldier seated on a trail leg wears a captured Soviet-style helmet and holds a 3.7cm shell.

(**Opposite, below**) The Wilhelmina bridge at Maastricht on the Mass River, destroyed by Dutch troops in the hope it would stop the German advance in May 1940. German troops of the *95. Inf. Div.* crossed the river in inflatable boats that were also used to ferry light anti-tank guns. Note the ammunition boxes slung over the gunshield.

(**Above**) A river-crossing drill in the mid-1930s using a raft made from two *Schlauchboot 34* boats, each with a cargo capacity of approximately 700kg. The raft was propelled either by paddles or a small outboard motor.

(**Opposite, above**) An anti-tank team of the *34. Inf.Div.*, which, following the Franco-German ceasefire, had been assigned to the occupation forces stationed in Boulogne-sur-Mer, where it undertook coastal defence responsibilities. The fate of Boulogne during the French campaign closely resembled that of Dunkirk. The city was held primarily by a mixed force of French and British troops, including elements of 20 Brigade of the Irish Guards. Despite heavy pressure, the defenders managed to hold the city long enough to facilitate the evacuation of the majority of British forces by sea.

(**Opposite, below**) During Operation Barbarossa, *Pak* guns were often transported on small rafts made of between two and four *Kleines Schlauchboot* pontoons lashed together with decking. The rafts were then towed across rivers by rope or motorboat.

Basic gun drill was conducted within the barracks, where soldiers were introduced to the fundamental elements of operating an anti-tank gun. This included learning crew roles, handling ammunition, loading and aiming procedures, misfire drills and synchronized movements required for efficient operation under combat conditions.

When on the move, 3.7cm *Pak* guns were fitted with a protective canvas cover to shield the weapon from dust, mud and moisture, helping to maintain operational readiness during long marches and harsh conditions.

Photo taken in autumn 1938, likely during the annexation of the Sudetenland. A *Pak* column is passed by eight-wheel armoured reconnaissance vehicles of the *2. Pz.Div.* The letter 'C' painted on the upper left corner of the gun's protective shield identifies the 3rd Section of the 1st Platoon of (likely) *Pz.Abw. Abt. 38*.

During pre-war firing drills, soldiers were permitted to wear side caps or garrison caps instead of helmets, as the primary focus of training was on improving aiming skills rather than full combat readiness.

Men of *Pz.Abw.Abt. 29* of the *29. Inf.Div.* (motorized) resting during a march break in mid-1938. Visible are the pre-war temporary tactical markings, including symbols identifying both the branch of service (triangle) and unit number. The white colour of the symbol indicates the 1st Company

The entire anti-tank battalion assembled in a barracks courtyard. At the front of the formation, the headquarters section is equipped only with staff cars and motorcycles.

A 3.7cm round has been fired and the gun barrel is at full recoil, as the *Geschützführer* (gun commander) observes the impact on the target.

Due to the small size of the *Pak* shells, crews could handle and reload the gun more easily in winter conditions, even while wearing thick gloves or mittens – an essential advantage in freezing temperatures.

The effectiveness of the 3.7cm *Pak* was decisively improved by the introduction of the *Stielgranate 41*. It was essentially a rod-type 'rifle grenade' loaded in the bore of the gun, with the finned sleeve sliding down outside the barrel. It could penetrate 150mm (6in) of amour at its effective range of 180 metres (200 yards) at 5° elevation

The *Leichttraktor* prototype armed with a 3.7cm *KwK* gun ended its service life as a monument at one of the German garrisons.

A *PzKpfw. III Ausf. A*, the first variant of the Panzer III series, equipped with a 3.7cm *KwK* gun, during field manoeuvres in the summer of 1938. The tactical number identifies this vehicle as belonging to the 4th Company, 1st Battalion, Panzer Regiment 2, 1st Panzer Division. A Panzer IV with a short-barrelled 7.5cm gun is visible in the background.

A long line of early *PzKpfw. III Ausf. B* and *A* tanks equipped with 3.7cm *KwK* guns. The photo was taken in mid-1938, most likely at the barracks of *Pz.Rgt. 1* in Erfurt.

The year of 1939 marked the transition of the *PzKpfw. III* from a series of experimental prototypes into the backbone of the *Panzerwaffe*. Early production models were armed with the 3.7cm *KwK* gun, which remained standard only until mid-1940, when it was gradually replaced by the more powerful 5cm *KwK*.

An early variant of the *PzKpfw. III* photographed in front of the *Panzer Kaserne* (tank barracks), with newly built large garages visible in the background. While artillery pieces were often stored outdoors under simple wooden shelters, all tanks were housed in enclosed brick-and-mortar garages to provide better protection and maintenance conditions.

A *PzKpfw. III*, most likely from *Pz.Rgt. 6* (1st Battalion, 4th Company, 3rd Platoon, Vehicle 3, based on the tactical number painted on the rhomboid plate). The photo was taken in 1938.

A *PzKpfw. III Ausf. C* photographed during a military display at the Zeppelin Field in Nuremberg, held as part of the Nazi Party rally (*Reichsparteitag*) in the autumn of 1938.

A PzKpfw. III from *Pz.Rgt. 2, 1. Pz.Div.* The photo was taken during the Polish campaign in 1939 near Klobuck. Note the early-war national markings in the form of a white cross.

A PzKpfw. III (222) and *PzKpfw. IV* of *Pz.Rgt. 1, 1. Pz.Div.*, during the Polish campaign. Both tanks have sustained damage, but not beyond repair. The recovery team has marked them with 'AFP' lettering, indicating they were qualified for restoration.

A *PzKpfw. III* armed with a 3.7cm *KwK* gun, still widely in service during the invasion of France in May 1940, as evidenced by this vehicle marked as part of the *10. Pz.Div.*

Two *PzKpfw. IIIs* cross a river in Romania preparing for the attack on the Balkans. All the waterproof covers, except on the commander´s cupola, have been fitted but do not have air-intake equipment.

A *PzKpfw. III* advances along a dusty rural road through the Greek mountains during the 1941 Balkans campaign. British Matilda IIs of 'W' Force (the Commonwealth expeditionary force sent to aid Greece) could withstand direct hits from 3.7cm guns, forcing the Germans to use flanking, high ground or air strikes (Stuka dive bombers) to knock them out.

Before the outbreak of the French campaign, some *Pak* guns were mounted on modified chassis of the *Krupp Protze Kfz. 69* artillery tractor. This adaptation significantly improved the mobility, allowing for quicker deployment and repositioning on the battlefield.

A *Pak* mounted on a *Kfz. 69* approaching the Marne River in mid-June 1940, just 80 miles east of Paris. The improvised self-propelled gun had a raised silhouette, making the vehicle more conspicuous and therefore easier for the enemy to detect and target, particularly in open terrain.

In June 1940, the *Gebirgs Jgd.Abt. 44* of the *1. Gebirgsjäger Div.* was equipped with a *Kfz. 69*, modified to mount a 3.7cm *Pak* on a special rotating frame. Two additional Krupp tractors were also converted into improvised self-propelled guns, these variants being further fitted with 8mm-thick armor plates shielding the driver's cab and engine hood.

A gathering of *Pz.Jgd.Abt. 97* (Bavaria) in the spring of 1941. The unit was equipped with captured French Renault UE vehicles, adapted to serve as gun and ammunition trailers. Only a few months later, the battalion was pressed into service on the Eastern Front.

At the start of Operation Barbarossa, some anti-tank battalions mounted their *Pak* guns directly on Renault UE tractors. The weapons were secured on two large wooden chocks and strapped down to prevent slipping, creating a simple yet effective form of improvised self-propelled anti-tank gun.

Pak guns were also issued to motorized infantry units, which were equipped with *Sd.Kfz. 251* half-tracks mounting the weapons. These vehicles were initially allocated to armoured battle groups spearheading the advance, under the command of generals such as Guderian and Kleist, many of them possessing the formation identification markings 'G' and 'K' respectively during the French campaign (1940) and the early phase of the Soviet campaign (1941).

An *Sd.Kfz. 251/10* with an unshielded 3.7cm *Pak*. The vehicle was used by a platoon commander. Such armed vehicles were used in mass scale only during the Russian campaign.

Sd.Kfz. 251/1 half-tracks fitted with *Pak* guns were designed primarily to suppress enemy positions and eliminate machine-gun nests. They also played an important role in protecting troop convoys equipped with unarmoured trucks, since by 1941 the number of available half-tracks was still insufficient to transport all *Panzergrenadiers*.

An *Sd.Kfz. 251/1* equipped with three external launch racks on each side for firing 28cm *Wurfgranate 42* (*WGr. 42*) rockets, and a *Pak* gun, photographed in Russia.

The summer field manoeuvres of *the 6. Inf.Div.* of the *Reichswehr* in 1934. At this time, the *Krupp Protze* – later to become the standard towing vehicle for anti-tank artillery – was still in development. *Mercedes-Benz 260* cars served as the primary towing vehicles for the *Pak* guns

A *Pak* gun towed by a *Horch 830R Kübelwagen*. Most of these early vehicles featured small side windows attached to the windshield frame, a typical design characteristic. Once the *Krupp Protze* became available for towing *Pak* guns, the *830R* vehicles were reassigned to signal units, where they were fitted with special carrying frames on the front mudguards, each capable of holding 300 metres of heavy field cable.

Assembly area of an anti-tank battalion equipped with *Mercedes-Benz 10/50 PS 'Stuttgart 260'* cars. More than 1,500 of this type were produced between 1929 and 1935, with as many as 1,507 built in the years 1934 and 1935 alone.

A *Pak* platoon equipped with *Kübelwagen* vehicles based on *the* Opel Super 6. Only a small series of these cars was produced specifically for the army; however, some military versions were converted from confiscated civilian models, of which more than 46,000 were manufactured between 1936 and 1938. This photo was likely taken in September 1939.

A *Pak* from *Pz.Abw.Abt. 14* towed by a *Kfz. 69* in 1937. This tractor type was fitted with four large ammunition lockers above the rear wheels, reducing the need for additional vehicles to haul ammunition trailers.

A *Pak* section on the move toward the Polish border in the final days of August 1939. Within hours, the Second World War would break out.

During the advance, especially while pursuing the enemy, *Pak* crews driving *Kfz. 69* tractors had to rely on canned food, as field kitchens could not keep up with the rapid movement. Combat rations were consumed during brief stops or even while on the move.

The *Sd.Kfz 251* of *Kp.15, Schutz.Rgt. 1, 1. Pz.Div.* used as a non-standard towing vehicle during the French campaign.

A Renault UE *Chenillette* advancing along the southern sector of the German offensive in the Soviet Union in the summer of 1941.

By the summer of 1941, all German motorized infantry divisions were equipped exclusively with tracked tractors (*Sd.Kfz. 10*) for towing *Pak* guns. This offered a significant advantage on the poor Soviet roads, as these vehicles, weighing around 3.5 tones, were comparable in weight to medium Soviet trucks and could safely cross smaller wooden bridges. (Bundesarchive Bild 101I-208-0038-15)

In unmotorized infantry divisions, *Pak* guns were even by mid-1941 towed by requisitioned civilian tractors. The vehicle shown here still carries its civilian registration beginning with 'IIB' (Berlin code), while the load box has been additionally marked with the military 'WH' (*Wehrmacht*) designation.

A *Kfz. 69* (*Horch 901* model), which was used as a towing vehicle for the 3.7cm *Pak*, beginning in 1940. This particular vehicle belonged to *Pz.Abw.Abt. 293, 293. Inf.Div.* Originally formed for the planned invasion of Britain, the division was ultimately assigned to the Russian campaign, where it was largely decimated by early 1943.

Chapter Four

Doctrine and Tactics

By the end of the 1930s, most countries based their anti-tank doctrine on experiences from the First World War and a few interwar conflicts, such as the Spanish Civil War and the Italian conquest of Ethiopia – where tanks had been used only on a limited scale. These engagements offered little indication of the large-scale, combined arms operations that would soon dominate the battlefield. Peacetime manoeuvres also failed to validate or challenge these doctrines effectively, as they rarely reflected the true dynamics of opposing armoured forces in combat.

The prevailing concept of anti-tank warfare at the time emphasized passive defence. Anti-tank guns were deployed to support infantry, who were tasked to separate enemy tanks from their supporting infantry and funnel them into pre-arranged 'kill zones' covered by multiple layers of anti-tank fire. This tactic was seen as effective because of the high mobility of tanks compared to the relatively static nature of infantry anti-tank operations.

The deployment of anti-tank weapons was closely tied to the positioning of the infantry units they supported, requiring a high degree of coordination. Defensive positions were often deeply dug to allow gun crews to remain protected, even from the crushing weight of passing tanks.

Anti-tank guns were typically organized into battalion-level formations, which at the time appeared adequate. Few armies – apart from Germany and the Soviet Union – anticipated the use of large, fast-moving armoured formations capable of breaking through front lines and exploiting deep into the enemy's rear. This concept remained largely theoretical – and in the case of Germany and the USSR, a closely guarded secret – until the invasion of Poland in September 1939, when both nations demonstrated the effectiveness of such tactics in practice.

More advanced doctrine introduced the concept of 'active anti-tank defence', which involved the flexible deployment of anti-tank weapons using motorized transport, and the creation of dedicated anti-tank reserves to support counterattacks. This approach was difficult to implement within traditional foot infantry formations, which lacked mobility and offensive capacity against armour. However, it became viable within motorized infantry and armoured divisions, particularly with the support of motorized anti-tank battalions.

In the years immediately preceding the Second World War, most German anti-tank units – known as *Panzerabwehr-Abteilungen* (Anti-Tank Battalions) within armoured, light and motorized infantry divisions – were trained in this active defence doctrine.

This approach, however, could only be achieved with a fully motorized anti-tank force, which was the case for the *Wehrmacht*.

Training of the *Pak* Gun Crew

The *Dienstunterrichtsbuch für die Panzerabwehr-Kompanie* (Instruction Manual for the Anti-Tank Company) outlined rigorous standards for training anti-tank formations, company commanders and each individual member of the gun crew.

The commander of a gun section was required to lead his team independently and therefore had to be a non-commissioned officer possessing strong personal qualities: a sense of responsibility, sound judgment, willpower and tactical acumen.

The gunner was expected to demonstrate specific character traits, notably keen alertness and the ability to act decisively and calmly under pressure to destroy enemy armoured vehicles.

The driver needed to be not only a disciplined soldier but also an expert technician, thoroughly familiar with his vehicle. High physical fitness was expected of all crew members.

Once individual roles were mastered with precision – allowing any member to substitute for another – training advanced to full-crew exercises focusing on coordination and teamwork.

Field exercises emphasized the rapid alignment and reorientation of the gun, swift occupation and abandonment of firing positions and efficient manoeuvring, particularly when towing with a vehicle. In typical terrain, two crew members pulled the gun; in more challenging conditions, the entire crew assisted. Ammunition handlers carried ammunition crates/containers at the rear, while the gun commander walked ahead, guiding the route.

At the firing position, the gun crew maintained full-sector observation: the gunner (No. 1) and loader (No. 2) observed the main sector; the first ammunition handler (No. 3) monitored the left and maintained visual contact with the commander; and the second handler (No. 4) covered the right and rear, maintaining communication with the platoon headquarters.

Aiming instruction prioritized firing at moving targets. Crews were trained to estimate both range and vehicle speed accurately – specifically at 300, 600 and 900 metres, and for tank speeds of 8, 16 and 24km/h. Once accurate direct fire was achieved, instruction expanded to engaging targets approaching from various directions and angles.

Tactical Principles of Anti-Tank Unit Deployment

According to pre-war regulations for anti-tank companies (*Panzerabwehreinheiten der Truppen*), organic anti-tank units (embedded within the troops) were responsible for protecting their own formations against enemy armour. In contrast to organic anti-tank units, the high-command reserve anti-tank units were organized as highly mobile reserve units to be deployed by higher tactical commands. Organic units typically operated by platoons, while reserve anti-tank formations were used as full units, particularly when facing a strong, layered armoured attack.

The smallest independent firing element was **the gun crew**, capable of autonomous action. Their firing positions were to be well-concealed, camouflaged, dug in and offer broad observation arcs to prevent surprise. If preparation time was insufficient, the crew used a 'waiting position' – a concealed standby location from which the gun could be quickly deployed to the firing line when needed.

Optimal firing positions were located behind terrain features such as ditches or minefields to protect the flanks and frontal arc. These positions required covered approaches, pre-planned withdrawal routes and separation from towing vehicles to reduce vulnerability. Key roads and enemy approach routes had to be under observation and, if possible, under fire.

Distances to likely enemy approach points had to be accurately determined up to 1,000 metres. Although the gun was expected to be ready to fire at this range, doctrine advised waiting until the enemy closed to 600 metres before opening fire. Following each engagement, the firing position had to be changed to avoid destruction by returning enemy forces.

Each independent **anti-tank platoon**, typically equipped with four *Pak* guns, was assigned a pioneer detachment and communication assets. The platoon commander generally positioned himself near the central gun to maintain oversight and command. Guns were deployed in a triangular pattern, with overlapping observation sectors, within a front no wider than 1,400 metres.

A dedicated command group conducted reconnaissance, security and communication. The commander held full authority over gun assignments and required excellent situational awareness, rapid decision-making and clear, concise orders. The platoon's supply train was to remain well-camouflaged and protected, ensuring uninterrupted ammunition delivery.

Companies operated either as whole units or by deploying individual platoons. Flank security was typically entrusted to the outermost platoons. Companies were trained to deploy from marching formation into firing positions rapidly and without confusion. Platoon positions overlapped to provide mutual support, and forward observers were placed on elevated ground to signal the approach of enemy tanks using lights or prearranged signals.

Platoon positions were designated on maps, or when possible, assigned in the field with terrain orientation. Readiness positions were selected far enough from firing positions to permit reaction time. On the march, flank protection was provided by platoon vehicles and motorcycle scouts. The company commander positioned himself near the front, behind the lead scouts, with the remainder of the unit following at proper intervals.

In the event of a sudden enemy encounter, companies were trained to transition into combat directly from march formation. Each platoon independently occupied the nearest available terrain, spacing itself for safety and maintaining fire coordination and flank coverage. If time allowed for organized deployment, the company could assume firing positions more rapidly and with greater cohesion. If caught by surprise, platoons deployed instinctively: the lead platoon engaged frontally, the centre platoon shifted right and the rear platoon moved left.

Standard operating procedures were instilled in all units to ensure seamless transitions from movement to combat, even under duress and without direct orders.

Examples of Combat Exercises for a Gun Section

Occupying a Position Within an Infantry Sector

A typical gun section was equipped with either two *Kfz. 12* vehicles (one for towing the gun, the other for the ammunition trailer) or a single *Kfz. 69 Krupp Protze* from 1937–38 onwards. Upon reaching the combat zone, the vehicles – under the direction of the lead driver – were hidden and camouflaged. The gun was then unlimbered and moved manually by the crew under the command of the gunner, while the section commander advanced ahead to reconnoitre and select a firing position.

Ammunition handlers carried two containers each. In difficult terrain, they could temporarily set down the crates to assist in moving the gun. If a concealed, dug-in position could not be found, the commander selected a 'waiting position' to conceal the weapon until needed.

The firing position had to allow engagement of targets up to 1,000 metres away, even though standard doctrine called for opening fire at 600 metres. At the commander's signal, the crew pushed the gun forward, set it up and camouflaged it, leaving only minimal openings for sighting. The immediate foreground was cleared of brush to improve visibility.

The commander initially observed the terrain, then passed this duty to the gunner, who adjusted the sights. The commander prepared a range sketch for the platoon leader, marking key reference points and probable enemy approaches at 300, 600 and 900 metres. The gunner received a duplicate of the sketch. Final checks included testing barrel traverse and elevation.

Once in firing position, the crew scanned all sectors while the commander coordinated with infantry on signalling methods for tank sightings, such as flashing a yellow light. Maintaining communication between the firing position and the vehicles was the driver's responsibility, responding to signals such as 'ammunition forward'.

Advancing Behind Infantry

Following a successful infantry advance, the *Pak* gun was withdrawn and moved manually in bounds 500–1,000 metres behind the line. Gunner and loader pulled the weapon using straps ('draught traces'), assisted by the ammunition handlers. Containers were carried or hung on the gun shield. The section commander scouted ahead, choosing the best route. Infantry were expected to assist if the terrain made gun movement difficult.

Occupying a new firing or waiting position followed the same procedure. In offensive operations, waiting positions were preferred due to time constraints. From these, the gun could be pushed into firing position upon alarm. Crews were expected to be combat-ready – with gun set and aimed – within two to four minutes of receiving the alert. If the terrain prevented vehicle movement, the towing vehicles remained at their original positions.

Engaging in a Firefight

When alerted to an approaching armoured threat – either by infantry or the observation service – the *Pak* gun was rapidly traversed to face the threat. On the commander's order, the gunner engaged the most visible tank, then shifted fire to the next closest after confirming a hit. This process continued until the threat was neutralized.

Throughout the engagement, the commander monitored effectiveness and ensured that the gun was not caught off guard by flanking movements. After repelling the attack – typically delivered in waves – the crew withdrew the gun to concealment and occupied a new, pre-prepared position. If the gun was rendered inoperable, the crew was expected to organize a defensive position and continue fighting using infantry tactics.

After each engagement, the gun position had to be changed without exception to avoid detection and destruction in a follow-up attack.

Combat-Proven Principles

While pre-war doctrine often called for engaging tanks at the maximum range of anti-tank weapons, it was learned early in the war that it was more effective to wait until tanks were within a few hundred metres, effectively springing an ambush. Rather than forewarning the advancing tanks with missed shots, which allowed them to take evasive manoeuvres, it was better to ensure hits at relatively close range.

Blitzkrieg tactics demonstrated that prepared positions for *Pak* guns were a rare luxury in fast-moving battles. Anti-tank guns were often hastily deployed on street corners or in roadside ditches, offering their crews only minimal protection.

During the French and North African campaigns, the Germans learned that panzer divisions, when structured as all-arms formations – including motorized, towed anti-tank gun battalions – could play an aggressive and decisive role in offensive operations. A common tactic involved tanks engaging the enemy and then pulling back deliberately, luring opposing armour onto prepared anti-tank gun positions. Once the enemy had been drawn in and disrupted, German armour would counterattack from the flanks, exploiting the weakened formation.

In North Africa, British forces often misattributed their tank losses to German tanks. In reality, the majority of tank casualties were inflicted by concealed or well-positioned German anti-tank guns, not by enemy armour.

By 1942/43, during the invasion of the Soviet Union, the nature of anti-tank warfare had changed in many respects. Armour was employed in both smaller-scale and mass operations. Self-propelled assault guns became more common on the battlefield. Conventional anti-tank guns were being upgraded to medium and large calibres, but improvements in tank design, armour, firepower and mobility often outpaced the deployment of adequate anti-tank guns. Quick fixes were common, with existing low-calibre guns being equipped with new ammunition types, as was the case for the 3.7cm *Pak*.

The *Stieilgranate 41*, capable of penetrating 150mm of armour with an effective range of 180–200 metres, attempted to extend the *Pak*'s usefulness, but the gun was so obsolete that it had to be replaced, which was easily done with the 5cm *Pak 38*, a scaled-up version of the 3.7cm gun, which was fielded in late 1940.

Even though the 3.7cm *Pak* became tactically outdated as an anti-tank weapon by 1942, it was far from useless. It found new life as a support weapon, training tool and defense weapon, especially in light or mobile units and secondary theatres.

(**Above and below**) Peacetime manoeuvres focused on developing effective methods to hide weapons from enemy observation, especially as the threat of aerial reconnaissance increased. The *Pak* gun crews would have been trained to use natural terrain features, such as trees, bushes and hills, to blend their positions into the environment.

The *L/45* calibre *Pak,* firing standard armour-piercing ammunition, could penetrate 29mm of armour at a range of 500 metres. While this was considered impressive in 1935, it proved highly inadequate by 1940, when *Paks* encountered French Char B1 heavy tanks with hull frontal armour of up to 60mm and turret front armour of 46mm.

An anti-tank company returning from field exercises to a newly built barracks. With the expansion and motorization of Germany's anti-tank units in the 1930s, many new facilities had to be constructed to accommodate vehicles, provide adequate storage for new equipment and offer improved living conditions for soldiers.

A 3.7cm *Pak* gun during a static live-fire exercise at the training range. The crew is supervised by an NCO, recording the results of the firing practice. Such meticulous record-keeping was a routine part of training, ensuring accuracy, discipline and evaluation of both the gun's performance and the crew's proficiency.

A clearly posed photograph of a *Pak* crew, likely taken as a commemorative keepsake rather than during training or combat. Such portraits were common in the interwar and early *Wehrmacht* period, as they allowed soldiers to proudly display their unit, weapons and equipment.

(**Above**) Field training represented the final stage of instruction for anti-tank gun crews, combining all previously learned elements into a coordinated exercise. Crews practiced deployment, concealment, rapid movement, target acquisition and live-fire drills under realistic conditions, ensuring they could function effectively as a team in combat.

(**Opposite, above**) During the preparation for opening fire, all crew members were expected to take a kneeling position. This served to minimize their visibility to the enemy and allowed for better coordination around the weapon. The gunner and loader remained closest to the breech, while the commander and ammunition handlers kept low but ready to move quickly.

(**Opposite, below**) Many field training exercises concluded with parades as the units returned to their home garrisons. These events not only demonstrated the discipline and professionalism of the troops but also served to strengthen the bond between the military and the local population. Crowds of inhabitants often lined the streets to welcome their soldiers, turning the return into a morale-boosting celebration that highlighted both civic pride and military readiness.

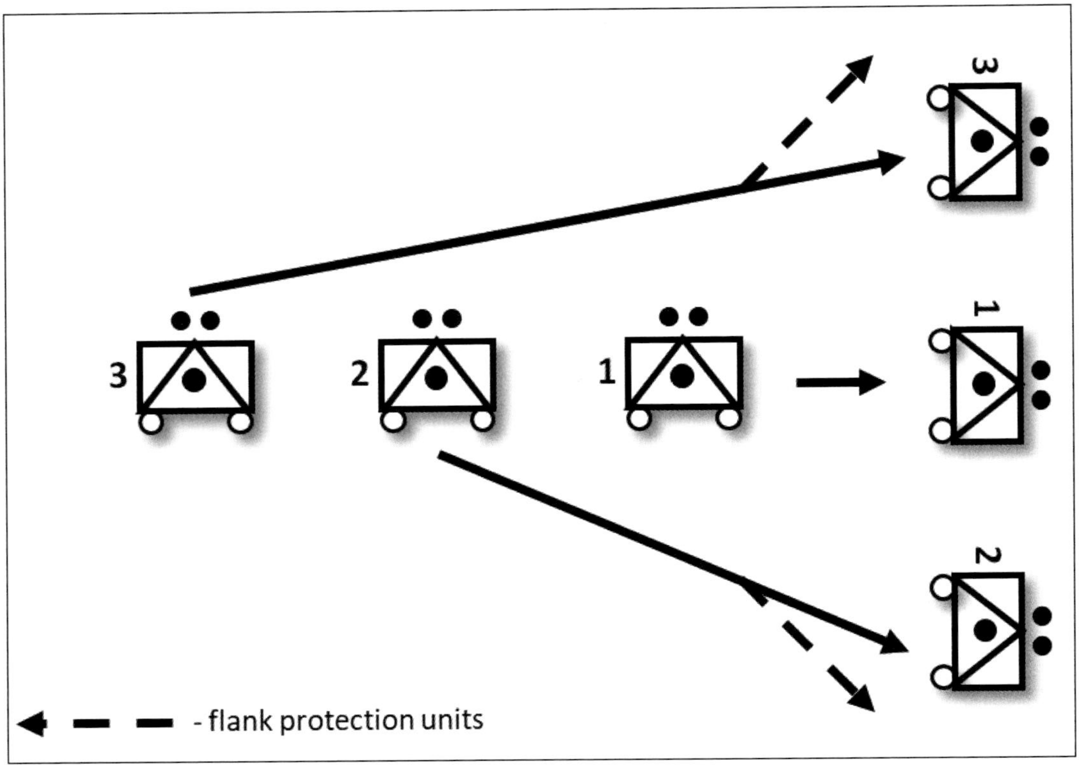

(Above) **Deployment of the AT Company from March Column to Platoon-Level Fire Positions**

(**Opposite, above**) **Manual Towing of the Gun to Firing Position** – *Crew Member Positions During Gun Movement*

Anti-Tank Gun in Firing Position – *Crew Positions and Assigned Observation Sectors*

Pak gun crew: (●) Section/Gun Commander – responsible for choosing positions, giving firing orders and coordinating with infantry; **(❶) Gunner** – Aimer/shooter, operates the sight and fires the gun; **(❷) Loader** – loads shells into the gun; **Ammunition Handlers (❸ & ❹)** – carries and supplies ammunition, helps move the gun and assists under fire.

(**Opposite, below**) Barrel cleaning under combat conditions was a critical task, ensuring that the *Pak* gun remained reliable and accurate. Even a small buildup of residue could reduce performance or risk a malfunction during battle.

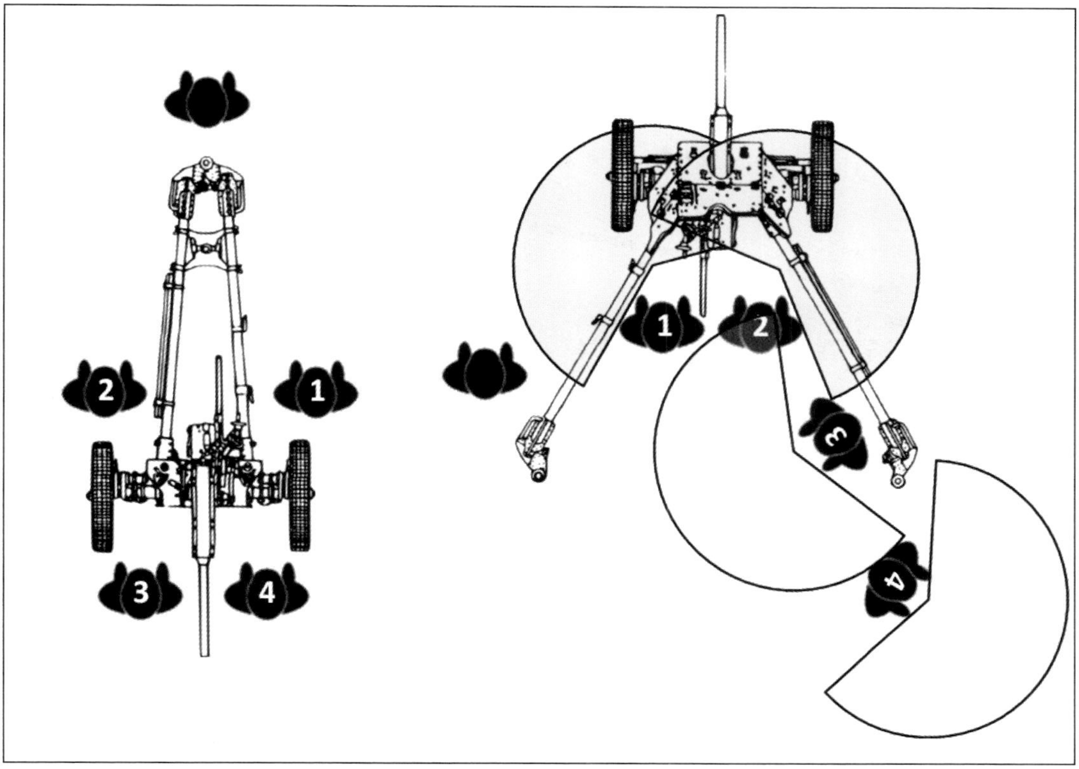

2
1
3
4
1
2
3
4

On 7 September 1939, a German anti-tank company halted briefly at the main square of Rawa Mazowiecka during its advance toward Warsaw. The guns and limbers are drawn up in formation, crews resting for only a moment before pushing forward.

The dark-grey painted protective shield of the *Pak* gun is seen here covered in a thick layer of dust and sand, a common problem for German equipment during the September 1939 campaign. The unusually hot and dry autumn created challenging conditions for machinery, but it also provided the advantage that many smaller rivers could be crossed by fords without the need for bridging equipment.

A *Pak* crew in firing position just below a Polish warning sign, which threatens penalties for crossing the boundary. Apparently, these restrictions did not apply to the German forces…

A *PzKpfw. I Ausf. A* cautiously advances down a snow-covered Norwegian road in April 1940, providing protection to the *Pak* crew moving alongside it. Armed with twin machine guns, the Panzer I offered suppressive fire in support of the anti-tank troops.

(**Above**) A *Pak* gun crew photographed after taking drastic measures to conceal and reinforce their weapon. The gun's splinter shield has been camouflaged and further reinforced with additional steel plating for extra protection.

(**Opposite, above**) A *Pak* gun positioned in front of a Maginot Line bunker along the French–German border, which stretched from Switzerland to Luxembourg. Although bypassed by the Germans in May 1940, some forts continued to resist until the armistice, eventually being forced to surrender.

(**Opposite, below**) During the advance into French cities, *Pak* guns were used to protect strategic points such as railway stations (as shown in this photo), bridges and major crossroads. While most *Pak* guns were provided with motorized tractors, heavier artillery – such as 105 mm howitzers – was still largely horse-drawn as late as May 1941.

In combat positions, *Pak* guns were often concealed with foliage or improvised coverings using *Zeltbahnen* – camouflaged, waterproofed personal rain capes, typically 1.7 × 1.7 metres in size, with grommets and straps – which could be combined to form a tent or small shelter for a group of men.

A commemorative photograph of an anti-tank section taken in April 1941, shortly after the capture of Yugoslavia.

A *Pak* gun crew positioned on a rocky hillside somewhere in Greece. The ammunition box is carefully prepared and placed next to the gun, ready for a firing mission. In the intense southern heat, soldiers often wore side caps instead of helmets and frequently rolled up their sleeves for comfort.

On the Eastern Front, greatcoats were highly valued as the harsh Russian winter set in. From November onward, the severe cold tested the endurance of all German soldiers. Note that *Pak* crews did not carry their rifles while operating the guns, focusing entirely on the weapon itself. However, their personal weapons were never far away, as every soldier still wore a bayonet attached to his belt, ready for close combat if needed.

German forces advancing toward Leningrad in the summer of 1941 faced major logistical difficulties. Roads were limited, often unpaved and frequently blocked or mined by retreating Soviet forces, who also staged ambushes along the routes.

To outflank Soviet positions blocking forest roads, *Pak* guns often had to be pushed into position by hand. Their size and weight made them one of the few weapons capable of silently reaching such position without alerting the enemy.

A German sniper, likely from a *Gebirgsjäger* (Mountain troop) unit, supporting the *Pak* team. While anti-tank guns targeted approaching light tanks, snipers focused on enemy officers, aiming to disrupt command and halt the advance.

By 1942, the most realistic tactic for the 3.7cm *Pak* crew was to aim for the tank's tracks in hope of immobilizing the vehicle, with the expectation that its crew might abandon the disabled tank under fire.

(**Above**) The *Pak* crew destroys hut after hut in a village defended by Soviet soldiers, setting the entire settlement on fire. Notably, one of the crew members is armed with a Soviet *PPSh* submachine gun instead of the standard *Mauser* carbine, a weapon highly valued by the Germans.

(**Opposite, above**) A *Pak* gun team from a German infantry division is ferried across a Soviet river in 1941, with an improvised towing vehicle – a captured Soviet truck

(**Opposite, below**) *Paks* could only penetrate the armour of Soviet T-34/76 tanks at a range of no more than 100–150 metres, requiring fire to be delivered from well-concealed positions that were hard to detect by Soviet infantry who were protecting the tanks. Setting up these positions in winter, in open fields, was time-consuming, often forcing gun crews to fire from exposed positions, which led to heavy casualties among them.

A *Pak* gun being repositioned by the anti-tank team of the *61. Inf.Div.* in the village of Gruzino on the Volkhov Front, January–February 1942.

Obsolete *Paks* in the East were frequently deployed to protect key strategic points (e.g. railway stations) that were critical for maintaining supply lines and troop movements.

By mid-1943, during the battles in the Kuban, rapid changes of firing positions were a life-saving necessity for German *Pak* crews. The Red Army in this sector possessed numerous guns of larger calibre and longer range, capable of destroying stationary *Pak* positions with deadly accuracy.

By the autumn of 1941, German forces were advancing into the hilly areas around Sevastopol in Crimea. *Pak* guns were used to suppress Soviet machine-gun nests, but they played no decisive role in the siege itself, which relied on the heaviest German artillery, including railway guns and super-heavy mortars.

By 1944, many outdated *Pak* guns were reassigned to reinforce fortifications along the West Wall and the Pomeranian Line, constructed before the outbreak of the war along the German–Polish border. It served as one of the last defensive barriers against the advancing Soviet forces, with even older weapons pressed into service to strengthen the *Pommernstellung*.

When circumstances permitted, *Pak* guns were dug in and connected to field telephones, which were used to coordinate all twelve guns of an anti-tank company. Such organized defensive positions were increasingly established from 1943 onward, as the *Wehrmacht* shifted from offensive operations to defending against Soviet attacks.

A *Pak* crew moving their gun into the ruins of Warsaw during the summer of 1944. Light anti-tank guns proved to be highly effective against Polish street fighters during the Warsaw Uprising, which lasted sixty-three days and resulted in the near-total destruction of the Polish capital. (Bundesarchiv_Bild_101I-695-0425-36)

Although rendered obsolete on the front lines by 1944, 3.7cm *Pak* guns continued to serve in second-line and training units up until the final months of the war. As the enemy advanced toward the German homeland in 1945, even these outdated weapons were pressed into combat, primarily used against soft-skinned vehicles in desperate defensive actions.

Chapter Five

Formations and Structure

The crew of the 3.7cm *Tak/Pak* gun consisted of five soldiers: the gun commander (*Geschützführer*, NCO – armed with a submachine gun, equipped with binoculars, compass, whistle, wire cutters and flashlight) oversaw the proper execution of all duties, ensured commands were followed and maintained safety protocols; the gunner (*Richtschütze*, Corporal, or *Oberschütze*, Senior Private – armed with a pistol, equipped with a shovel and a strap for pulling the gun); the loader (*Ladeschütze* – armed with a pistol, equipped with a shovel and a strap for pulling the gun);, and two ammunition handlers (*Munitionschütze*), each with a rifle and two ammunition canisters containing twelve rounds each. The crew was supplemented by the driver (*Fahrer/Schütze*, Private) of the vehicle transporting the entire gun crew, ammunition and towing the gun itself. Typically, the vehicle was a medium-sized off-road car with a towing hook, such as the *Kfz. 12*, or a light artillery tractor, *Kfz. 69*.

With the introduction of the *Tak* guns in 1930, they were initially assigned to seven infantry divisions, each consisting of three infantry regiments, a field artillery regiment, pioneer and signal units and one transport battalion (horse-drawn and motorized equipped) called *Kraftfahr Abteilung*, or in short *Kf.Abt.*

Each infantry regiment comprised three battalions (*Abtailungen I/II/III*), each with four to six companies. Company No. 5 (*5.Kp/Abt.II*) was equipped with heavy support weapons including heavy machine guns and light anti-tank guns.

Initially, due to shortages of *Tak* guns, some of these units were equipped with wooden-barrel (*Holzgeschutz*) anti-tank guns. The wooden barrels were quite massive, suggesting a larger calibre than 3.7cm. Units armed with these makeshift guns were referred to as *Holzgeschutzzug*.

While the Motorized Transport Battalions (*Kf.Abtn*) were formally responsible for transportation duties, starting from 1930 they began training to serve as motorized units capable of operating as tank and anti-tank formations, thus were equipped with 3.7cm anti-tank guns.

By October 1934, the seven Motorized Transport Battalions (*Kf.Abt. 1–7*). had been reorganized into fourteen Motorized Tank-Defence Battalions (*Kampfwagen Abwehr Abteilungen*, in short *Kw.Abw.Abt.*) and seven Motorized Reconnaissance Battalions (*Aufklärungs Abteilungen*, *Aufkl.Abt*), which retained the designation *Kf.Abt.* as a deception measure. They were assigned to two Motorized Combat Brigades (*Kraftfahr Kampfbrigaden*), each consisting of six *Kw.Abw.Abt.* and three *Aufkl.Abt.*, under the command of the *Kraftfahr Kavallerie Korps*. Furthermore, for East Prussia, an additional *Kraftfahr Kampf Rgt. 1* was formed under the command of *Kf.Kdo. Königsberg*.

The *Kw.Abw.Abtn* were stationed as follows: in Military District I (WK I) in Allenstein (*Kw.Abw.Abt. No. 11*) and Osterode (*No. 21*); in WK II in Munsterlager (*No. 2*) and Schwerin (*No. 12*); in WK III in Döberitz (*No. 3*) and Neuhammer (*No. 8*); in WK IV in Magdeburg (*Nos 4* and *13*); in WK V in Kassel (*No. 9*) and Würzburg (*No. 15*); in WK VI in Hamm (*No. 16*) and Hannover (*No. 19*); and in WK VII in Straubing (*No. 10*) and Würzburg (*No. 17*).

By 1934–35, 3.7cm anti-tank guns (*Taks/Paks*) were issued to infantry regiments and organized into Tank-Defence Companies (*Panzer Abwehr Kompanien*, abbreviated *Pz.Abw.Kp.*), each equipped with nine guns and an adequate number of motor vehicles for towing and logistics. Since each infantry division consisted of three regiments, a full division fielded a total of twenty-seven anti-tank guns deployed forward of the front line.

In addition to these regimental-level units, every infantry division also included a Tank-Defence Battalion (*Panzer Abwehr Abteilung, Pz.Abw.Abt.*) made up of three companies, each with nine 3.7cm *Taks/Paks*, totalling another twenty-seven guns. These were intended to engage enemy armoured formations head-on or to cover vulnerable flanks.

Furthermore, the divisional reconnaissance battalion (*Aufklärungs Abteilung, Aufkl.Abt.*) typically included one anti-tank platoon (*Zug*) equipped with three 3.7cm *Paks*, further extending the division's anti-tank capabilities.

Panzer Divisions

With the formation of the first three German armoured divisions in October 1935, three anti-tank battalions (formerly *Kw.Abw.Abtn*) were integrated into these units as *Pz.Abw.Abt. 37, 38* and *39*, each equipped with twenty-seven *Pak* guns.

In addition to the divisional-level *Pz.Abw.Abtn.*, *Pak* guns were assigned to the two Motorized Infantry Brigades (*Schützen Brigaden*) within each panzer division. Each brigade received two platoons, with three *Pak* guns per platoon.

Further anti-tank support was provided by assigning one platoon of three *Pak* guns to both the divisional reconnaissance battalion (*Aufkl.Abt.*) and the motorcycle rifle battalion (*Kradschützen Abt.*).

By 1939, a Tank-Defence Battalion of an armoured division had its final wartime established strength of 519 soldiers, with its armament including three companies equipped with 3.7cm *Pak* anti-tank guns (thirty-six in total). Its vehicle inventory consisted of sixty-three motorcycles, twenty-nine staff cars and sixty-eight lorries, with the primary tow vehicle being a light six-wheeled lorry like the *Krupp Protze*. Each anti-tank company was structured with twelve *Pak* guns, organized into four guns per platoon.

Interestingly, in September 1939, *Pz.Abw.Abt. 79* of the *4. Panzer Division* was still missing one of its three *Pak* companies, while *Pz.Abw.Abt. 53* of the *5. Panzer Division* was lacking its headquarters unit (*Stab*). However, the *1., 3.* and *4. Panzer Divisions' Panzer Abwehr Abteilung* each had a *Flak* (anti-aircraft) *schwere MG Kompanie* (mot.) attached, totalling 189 personnel (six officers, twenty-nine NCOs and 156 other ranks), equipped with twelve towed 20mm *Flak 30* or *Flak 38* guns.

On top of the thirty-six *Paks* of the Panzer Tank-Defence Battalion, the typical panzer division had an additional eighteen guns in its Rifle Regiments (*Schützen Regiment*).

Other divisions

An infantry division had a *Pz.Abw.Abt.* of thirty-six 3.7cm *Pak* guns, while its three infantry regiments had a total of another thirty-six *Paks* at their disposal. All the light divisions (*leichte Div.*), mountain divisions (*Gebirgs Div.*) and motorized infantry divisions were similarly equipped.

Each of them included a Tank-Defence Battalion organized into three companies (3 x 12 *Paks*) plus tank-defence within the rifle regiment of twelve guns distributed among four platoons. In total, a motorized division was supposed to have seventy-two 3.7cm anti-tank guns according to its establishment. However, actual numbers in units were lower.

During the first wartime campaign in Poland, the organization of anti-tank units – both at the divisional level and within companies attached to infantry battalions – proved to be effective. The number of anti-tank guns deployed at the front was considered sufficient. However, ahead of the French campaign, on 16 March 1940, the Tank-Defence Battalions were officially renamed Tank Hunter Battalions (*Panzerjäger Abteilung, Pz.Jgd.Abt.*) to emphasize a more aggressive, mobile approach, in line with evolving German doctrines such as active defence and combined arms tactics.

The *Pz.Jgd.Abt.* was a fully motorized unit, comprising a staff with an intelligence platoon, three companies (each equipped with twelve *Pak* guns) and supply trains. It had a total strength of 550 officers, non-commissioned officers and enlisted men. The battalion was still armed with thirty-six 3.7cm anti-tank guns and eighteen light machine guns, and equipped with 114 motorized vehicles and forty-five motorcycles.

A *Panzerjäger Kompanie* included a company commander (*Oberleutnant*) with a company troop of thirteen men using cars and motorcycles, four platoons (I and II led by a *Leutnant*, III by a *Feldwebel* and IV by an *Unteroffizier*), a combat supply train with one car and one large field kitchen, and three trucks. The company was armed with twelve anti-tank guns, four light machine guns, eighty-six rifles, four machine pistols and ten pistols.

Each anti-tank platoon (*Zug*) consisted of a platoon leader with a headquarters troop of seven men, three 3.7cm *Pak* guns with six men each and one light machine-gun troop with three men.

Additionally, each of the three infantry regiments within an infantry division had its own *Panzerjäger Kompanie No. 14* equipped with twelve *Pak* guns. Furthermore, one three-gun platoon was incorporated into the reconnaissance battalion of the division.

Although an infantry division, according to its establishment, should have had up to seventy-five 3.7cm *Pak* guns, the actual number of anti-tank guns in many units was much lower. The later the mobilization *Welle* (i.e., the later the division was formed during the war), the weaker was its equipment.

Nine divisions from the 13th *Welle* and eight from the 14th wave possessed only twenty-one anti-tank guns each. Fifteen divisions from the 15th *Welle* had no organic anti-tank artillery at all! The 13th to 15th wave divisions had such limited combat capability that they were used only for coastal defence, as well as for occupation duties in France, Poland, Norway and the Balkans. Four security brigades/divisions of the 16th *Welle* were formed for guard duty in occupied territories of the USSR and rear areas of the front, each intended to receive a company of anti-tank guns (twelve guns, often captured artillery). Ultimately, they did not receive *Pak* guns.

Additionally, four light infantry divisions (*leichte infanterie divisionen*) were formed for operations in mountainous and wooded areas of the Soviet Union. Each division comprised two infantry regiments. Within each regiment, anti-tank defence was strengthened by an organic motorized *Panzerjägerkompanie*, equipped with 3.7cm *Pak* anti-tank guns.

At divisional level, the infantry could also rely on further dedicated anti-tank assets. The divisional reconnaissance battalion included its own small motorized *Pak* section (three 3.7cm guns), while the divisional *Panzerjäger Abt.* fielded two motorized companies with a combined establishment of twenty 3.7cm *Pak* guns, supplemented by four heavy anti-tank rifles (*sPzB 41*). In total, this gave the division an anti-tank component of forty-four *Pak* guns and four *sPzB 41*s.

For towing duties, captured French Renault UE tractors were frequently employed.

Independent Anti-Tank Battalions

After the Polish campaign, several independent anti-tank/tank-hunter battalions were organized, each equipped with thirty-six 3.7cm *Pak* guns and not assigned to any specific division. These units were initially designated as follows: 543rd (Herford), 559th (Stuttgart), 561st (Berlin-Dahlem), 563rd (Dresden) and 611th *Pz.Abw.Abtn.* (Schlawe).

In the spring of 1940, all were re-designated as *Pz.Jgd.Abtn.* They remained independent until January 1941, when *Pz.Jgd.Abt.* 543 was incorporated into the *3. Panzer Division.*

In May 1942, *Pz.Jgd.Abt. 559* was equipped with Marder self-propelled anti-tank guns, and by April 1944, it was rearmed with Jagdpanther tank destroyers.

In October 1942, *Pz.Jgd.Abt. 561* also received Marder SPGs. In January 1943, *Pz.Jgd.Abt. 563* was issued towed 7.5cm *Pak 40* anti-tank guns. *Pz.Jgd.Abt. 611* received Marder SPGs in the spring of 1942. Before the invasion of the Soviet Union in June 1941, three more independent *Panzerjäger* battalions were formed, compensating for the earlier loss of *Pz.Jgd.Abt. 543* to the *3. Pz.Div.* These new units were the 463rd, formed in Norway (1942) and later incorporated into the 199th Infantry Division, the 652nd, formed in Brunswick and disbanded in April 1942, and the 654th, formed in Bremen and re-equipped with 7.5cm *Pak 40* guns by November 1942.

Before the start of the Russia campaign, on 21 April 1941, some infantry divisions still did not have full staffing in their anti-tank subunits. The 7th and 8th *Welle* formations, as well as the 12th *Welle* ones, had only two companies in their divisional anti-tank battalions (twenty-four instead of thirty-six guns). However, *Pak* guns were rushed to equip the newly formed SS divisions before the attack on the Soviet Union.

(**Above and below**) An interesting photograph of dummy *Pak* guns (*Holzgeschütz*), with wooden barrels, used by the *Reichswehr* until genuine *Pak* guns were available. A breech and barrel removed from a *Mauser 98k* rifle was mounted inside the wooden barrel, chosen for its ballistic characteristics, which allowed it to simulate the trajectory of a 3.7cm anti-tank gun shell.

(**Above**) By 1935, all motorized anti-tank companies were equipped with twelve guns, each supported by a motorized ammunition trailer, requiring twenty-four towing vehicles in total. The photograph shows the rear of an anti-tank column from *Pz.Abw.Abt. 41* during field exercises in Bavaria's Allgäu region. This area offered a mix of rolling hills, forests, river valleys and alpine foothills. This gave anti-tank units the chance to practice both movement on open ground and deployment in more restrictive, broken terrain.

(**Opposite, above**) An anti-tank platoon from *Kf.Abt. 4* (*Reichswehr* Motorized Transport Battalion) of the *4. Inf.Div.* in the main square in Bad Muskau, Upper Silesia, in July 1934. The guns were accompanied by light armoured cars of the *Kfz. 13/14* type.

(**Opposite, below**) *Pak* guns, likely from *Kp.3/Pz.Abw.Abt. 38, 2. Pz.Div.*, in October 1935. The gun's front shield is marked with a white triangle bearing the number '12', indicating it was the last gun in the 3rd Platoon.

RW 4869
RW 4870

The five-man crew of a 3.7cm *Pak* is ready to open fire on enemy armour advancing across an open landscape, but attacking a French tank at a range of more than 300 metres was futile.

The *Krupp Protze* (*Kfz. 69*) was powered by a 3,308cc Krupp M304 air-cooled horizontally opposed four-cylinder petrol engine. These vehicles carried six crew members, along with 180 AP and thirty-six HE rounds.

Possibly *Kp.1/Pz.Abw.Abt. 28* during a public presentation in the barracks courtyard at Breslau, in late autumn 1938. The timing – shortly after the Sudeten crisis – suggests this event may have been intended both as a display of strength and a morale exercise, underscoring the role of anti-tank troops within the rapidly expanding *Wehrmacht*.

The combination of a *Pak* and a *Kfz. 69* presented a versatile and effective anti-tank capability. The mobility was essential for countering enemy armour and supporting infantry operations.

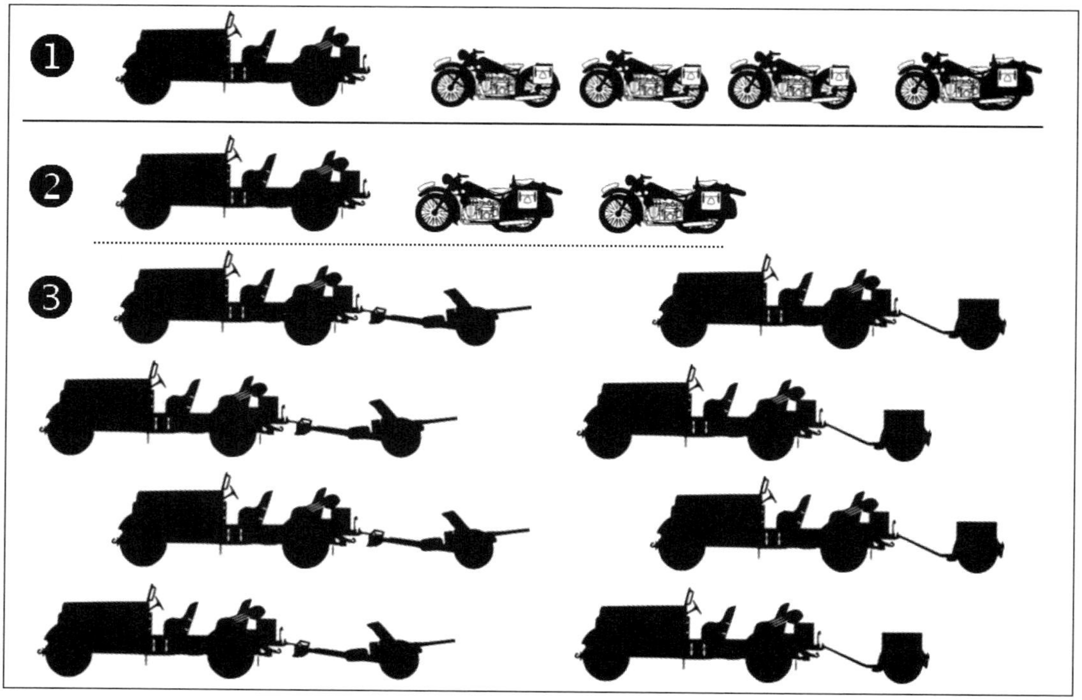

(Above) **AT Company '**'B'/**Panzer Abwehr Kompanie (mot.)**

OOB – Ausf. 'B' KStN 1143 1.10.1937

AT HQ Section/Troops: four motorcycles plus staff car (*Kfz. 12*) ❷ **1st AT Platoon: 1st Platoon (Plt.)** HQ Section: two motorcycles with sidecar and staff car (*Kfz. 15*); **1st AT Platoon:** four *Pak* 3.7cm guns, four *Kfz. 12* tractors, four *Kfz. 12* towing ammunition trailers. ❹❺ **2nd and 3rd Platoons, same as 1st Platoon; Company Trains: three** motorcycles, staff car, one light truck, three medium (3-ton) trucks.

(Opposite, above) AT Battalion 'B'/Panzer Abwehr Abteilung (mot.)/Pz.Abw.Abt. (mot.)

(Opposite, below) AT Company 'A'/Panzer Abwehr Kompanie (mot.)

OOB – Ausf. 'A' KStN 1141 1.10.1937

❶ **AT HQ Section/Troops:** four motorcycles plus staff car (*Kfz. 12*) ❷ **1st AT Platoon: 1st Platoon** HQ Section: two motorcycles with sidecar and staff car (*Kfz. 15*); **1st AT Platoon:** three *Pak* 3.7cm guns, three *Kfz. 69* (tractors, four *Kfz. 69* towing ammunition trailers. ❹❺ **2nd and 3rd Platoons, same as 1st Platoon; Company Trains:** three motorcycles, staff car, one light truck, three medium (3-ton) trucks.

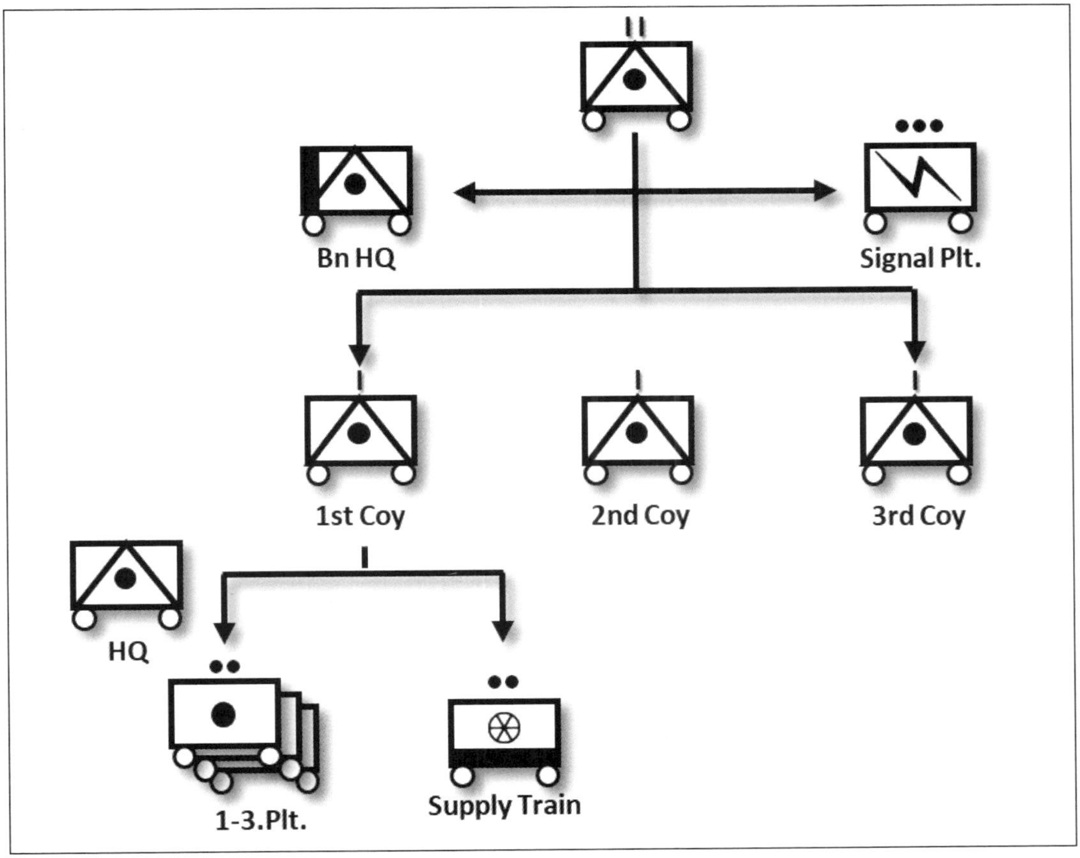
II
Bn HQ
Signal Plt.
I
I
I
1st Coy
2nd Coy
3rd Coy
I
HQ
1-3.Plt.
Supply Train

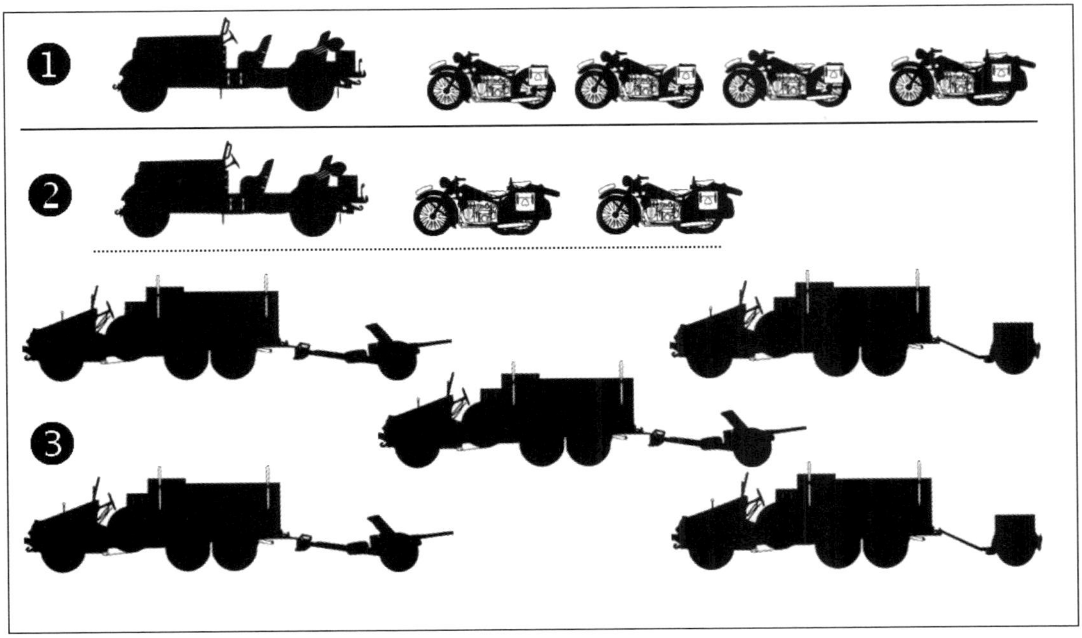
1
2
3

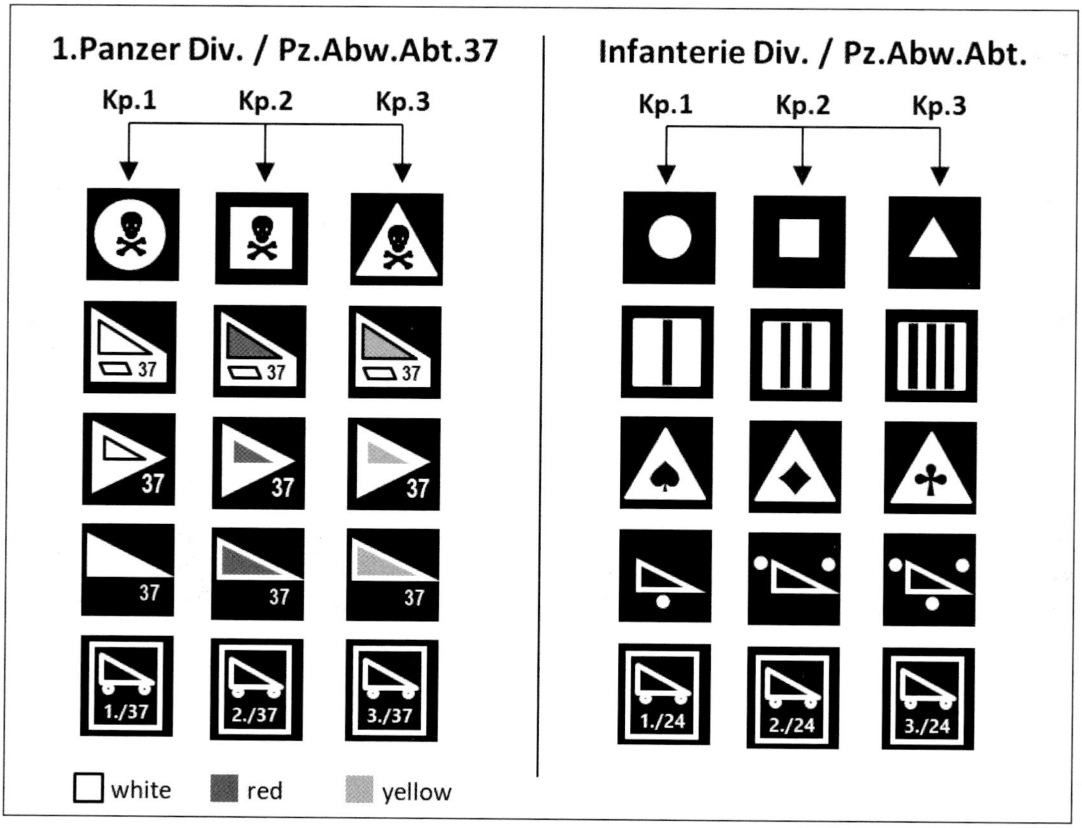

Evolution of AT Battalions Tactical Markings (09.1935–08.1939)

A presentation of major German artillery types in October 1937 at the Kempten Barracks of Artillery Regiment 27. Featured weapons included the 3.7cm *Pak* anti-tank gun, the 10.5cm *le.FH 18* field howitzer and the 7.5cm *le.IG 18* infantry support gun.

Pak gun in firing position, with a close-up view of its breech block, clearly showing the left-side trigger and the elevating handwheel.

Elements of a 1939-style *Pz.Abw. Kompanie* on the parade ground of their garrison. A 3.7cm *Pak* would be towed by a civilian-type 4x2 motor vehicle, with another towing a *Sonderanhänger* (*SdAnh.*, special purpose trailer) *32* carrying ammunition.

AT HQ section/troops, equipped only with staff cars and motorcycles, were responsible for command and control, reconnaissance and intelligence, logistics and supply, medical support and administrative duties – ensuring the unit's operational effectiveness.

A column of *Kfz. 69* vehicles towing *Pak* guns in the mountain region of the Lower Alps, spring 1939. The final months of peace were dedicated to intensive training, improving coordination and preparing all elements of the unit for future combat operations.

Preparations for a parade commemorating Adolf Hitler's birthday in April 1939. One of the key formations showcased was the *1. Pz.Div.*, including its divisional anti-tank battalion, *Pz.Abw.Abt. 37*.

Pre-war field exercise (note the red 'friend-or-foe' bands on the helmets). While on the move, guns were to be protected by a special canvas cover, or at least by a muzzle cap, clearly visible in the photo.

(**Above**) Gathering of *Pz.Abw.Kp.14/Inf.Rgt. 102* at the Chemnitz Barracks in August 1939. On 1 September 1939, the regiment, along with the rest of the *24. Inf.Div.*, crossed the Polish border at Neumittelwalde in Silesia and advanced over the Prosna and Warta rivers to the Bzura. There, it engaged in heavy fighting with Polish units attempting to break out from the Kutno pocket.

(**Opposite, above**) Winter field exercises were always the most challenging, with the risk of tractors becoming stuck in deep snow or mud. Therefore, crews were equipped with extra shovels and pickaxes, sometimes fixed to the front bumper. Such additional equipment was permanently assigned to *Pak* units while on the Eastern Front during autumn operations.

(**Opposite, below**) In traditional foot infantry regiments, anti-tank guns were primarily employed in a static, reactive role – forming part of a layered defensive line to blunt or halt enemy armoured assaults. In contrast, within motorized and mechanized formations, these guns were used more flexibly and aggressively.

(**Above**) *Pak* guns were frequently employed as road-blocking weapons to deny enemy movement and control key approach routes. Positioned at chokepoints, crossroads or concealed along expected lines of advance, these guns proved highly effective against lightly armoured reconnaissance vehicles and armoured cars.

(**Opposite, above**) By 1942, *Pak* guns had become ineffective, but some were still retained, even in panzer divisions. However, at this stage, primary anti-tank weaponry consisted of Marder III tank destroyers, built on a tracked chassis of *PzKpfw. 38(t)* and armed with a captured Soviet 7.62cm gun mounted in an open-topped fighting compartment.

(**Opposite, below**) The *Sd.Kfz. 10*, a light half-track artillery tractor and personnel carrier introduced in 1937, was originally designed to tow light anti-tank and anti-aircraft weapons. After the French campaign, an unknown number were converted into self-propelled guns by mounting the 3.7cm *Pak* on the vehicle's chassis.

This particular *Sd.Kfz. 10* of Heinz Guderian's *Panzergruppe 2* is seen advancing in the Minsk area during the summer of 1941, where it took part in the encirclement battle that trapped over 300,000 Soviet troops.

Mounting the 3.7cm *Pak* on the *Sd.Kfz. 10* greatly improved the gun's mobility, but it also increased its silhouette, making concealment more difficult. The vehicle's thin armour offered only minimal protection to its crew, leaving them vulnerable to small-arms fire and artillery fragments. Nevertheless, these makeshift self-propelled guns filled a crucial gap in mobile anti-tank defence during the opening months of Operation Barbarossa.

Chapter Six

Small Gun, Big War

The Spanish Campaign

The 3.7cm *Pak* gun saw its combat debut during the Spanish Civil War in 1936, having been delivered alongside German instructors serving with *Panzergruppe Drohne*. Of the first twenty-four guns sent to Seville, thirteen were used to form a Motorized Anti-Tank Company, initially attached to the German armoured group. This unit was later transferred to the Spanish Foreign Legion's Tank Regiment (*Bandera de Carros de Combate de la Legión*).

These early deliveries laid the foundation for what would become the Francoist Anti-Tank Batteries and several independent sections, which eventually received around 290 guns over the course of the conflict, delivered in successive shipments from Germany.

The first significant encounters between German *Pak* guns and Soviet-supplied Republican tanks occurred in October and November 1936 around Madrid. These engagements demonstrated the lethal effectiveness of a well-concealed *Pak* against lightly armoured Soviet T-26 tanks. The Republicans lost sixteen tanks outright and suffered damage to another thirty-six, mostly due to anti-tank fire.

Conversely, the T-26s, which were armed with the Soviet 45mm gun (an enlarged version of the German *Pak*), proved devastating against the German-supplied Panzer Is used by the Nationalists. Republican tank crews quickly realized they could engage enemy armour at ranges of 1–1.5km, staying well beyond the effective range of the 3.7cm *Pak*, which struggled beyond 900 metres.

Nonetheless, the *Pak* proved effective when deployed strategically, as Republican tanks often advanced in small formations to support slow-moving infantry, which frequently lagged behind, leaving tanks exposed. Even when infantry managed to spot the anti-tank positions, they lacked reliable means of communicating with the tank crews to warn them of threats.

The Republican offensive at Brunete in July 1937 became a key case study for Western military observers, thanks to heavy media coverage.

The operation aimed to relieve Madrid through a wide flanking manoeuvre intended to encircle Nationalist forces from the west and southeast. On 6 July, the XVIII Corps' attack on Villanueva de la Cañada began poorly. A battalion of T-26 tanks advanced across open ground, with the 34th Division's infantry following behind. The tanks were stopped roughly 500–600 metres from the town by two well-hidden 3.7cm *Pak* guns and a pair of field guns. One of the *Paks*, cleverly positioned in a church steeple, was credited with knocking out a dozen tanks on its own.

Despite fierce resistance, Republican forces managed to advance up to 20km into Nationalist-held territory. However, their progress was slow and lacked momentum, which allowed the defenders time to regroup and re-establish a solid front line.

The inability of tanks to break through enemy anti-tank screens at Brunete was widely interpreted by Western analysts as a sign of the tank's limitations. Prominent theorists such as British armour advocate B.H. Liddell Hart even began to question the tank's ability to restore mobility to warfare. Others, however, argued that the tanks had simply been misused, and cautioned against drawing sweeping conclusions from the Spanish experience.

While opinions on tank effectiveness in Spain were mixed, the performance of the 3.7cm *Pak* was widely praised. It became a benchmark for anti-tank guns among European armies, as well as those in the Soviet Union and the United States. Its capability to destroy enemy tanks, machine-gun nests and fortified positions using high-explosive fragmentation shells highlighted its versatility and solidified its reputation as a key battlefield asset.

The Polish Campaign

The next test for the *Pak* guns was the Polish campaign of 1939, during which anti-tank guns rarely had the opportunity to fight against Polish armoured vehicles. Nevertheless, when such engagements did occur, the *Pak* guns confirmed their effectiveness.

For instance, on 3 September 1939, near Lake Mełno in Pomerania, the 37cm guns from the German *228. Inf.Div.* destroyed or damaged eight out of twelve attacking TK-3 tankettes belonging to the 81st Reconnaissance Tank Company.

There were more examples of German anti-tank guns' effectiveness in combat against Polish armoured fighting vehicles. On the morning of 12 September, Polish 7TP tanks from the Light Tank Company of the Warsaw Defence Command, attacking alongside infantry towards Okęcie, initially achieved significant success, forcing one of the battalions of the German *31. Inf.Div.* to retreat. However, later on, the threatened sector saw the German *4. Pz.Div.* deploy an infantry battalion and an anti-tank squadron with *Pak* guns. Under their fire, the Polish tank attack was repelled.

During the September campaign, *Paks* were mainly used to support infantry, firing at targets such as machine-gun nests, mortar positions, field fortifications (including in the battles at Wizna, where they were used to shell fixed defences) and various types of mechanical vehicles.

Over the course of the fighting in Poland, the *Wehrmacht* and *SS* units collectively lost a total of ninety-seven 3.7cm calibre guns, with these types firing as many as 1,793,000 projectiles during the three weeks of fighting (based on *Abschukmeldung* from 25 October 1939), mainly high-explosive shells, constituting 12 per cent of the ammunition stockpile for 3.7cm guns at that time.

Certain weaknesses of the *Pak* gun became apparent during the Polish campaign. The most significant were weak axles and other suspension elements, which had a tendency to break when towing the guns across rough terrain or on unpaved roads. As a result of this experience, improvements to the axles were gradually implemented in production and also retroactively applied to previously manufactured guns.

It is worth noting that during the Polish campaign, most divisional anti-tank units were not yet fully developed to their authorized strength. The primary reason for this was not a lack of *Pak* guns, but rather a shortage of tractors to tow them.

The French Campaign

During the French campaign conducted in May and June 1940, it became evident that *Pak* guns were losing their effectiveness in combat against modern tanks used by the Allies. They struggled to destroy even infantry support tanks like the Renault R-35, which had armour ranging from 32–45mm thick. Very often, such vehicles could only be destroyed at distances of around 300 metres, unless attacked from the side, in which case this distance doubled.

The 3.7 cm shells proved completely ineffective against well-armoured British Matilda II tanks and French medium tanks like the B 1bis and Somua S-35, which had armour thicknesses reaching up to 78mm. Heavy losses were sometimes suffered; for example, the *57. Inf.Div.* lost twelve *Pak* guns destroyed by enemy shells or run over during seven days of fighting in the Abbeville bridgehead area.

When a French armoured formation under General de Gaulle had probed the Panzer flanks with two improvised raids carried out with inadequate forces near Arras on 20 and 21 May, *Paks* once again failed to live up to its expectations. The Allied counterattack severely dented German confidence, and at one point looked as though it might halt further German progress altogether. It did demonstrate that the 3.7cm gun was of little use against the armour of Matilda tanks, as proved by the account by Brigadier Douglas Pratt:

> 'One Matilda had fourteen direct hits from one of [...] 37mm guns and it had no harmful effect, just gouged out a bit of armour! The main opposition came from his field guns, some of which fired over open sights.'

The German commander in this action was *Generalmajor* Erwin Rommel, and his account confirms the inadequacy of the 3.7cm *Pak* gun:

> 'The anti-tank guns which we quickly deployed showed themselves to be far too light to be effective against the heavily armoured British tanks, and the majority of them were put out of action by gunfire, together with their crews, and then over-run by the enemy tanks. Many of our vehicles were burnt out. [...] Finally the divisional artillery and the 88mm anti-aircraft batteries succeeded in bringing the enemy armour to a halt south of the line Beaurains–Agny. Twenty-eight enemy tanks were destroyed by the artillery alone, while the anti-aircraft guns accounted for one heavy and seven light.'

Under favourable circumstances, however, the *Pak* remained a very effective weapon. An example of this was the battle fought by the German *Inf.Rgt. 102 (24. Inf.Div.)* on 23 May 1940, during a French counterattack south of Sedan. *Leutnant* Müller of the *14. Kompanie, Inf.Rgt. 102*, directing fire from two *Paks*, managed to destroy five enemy tanks and forced another six to retreat.

Even more effective was *Gefreiter* Brinkforth of the *14. Kompanie (Inf. Rgt. 25)* at the Abbeville bridgehead, where he destroyed eleven out of twenty attacking tanks in just twenty minutes, thereby thwarting an assault against his *2. Inf.Div.* Brinkforth became the first German enlisted anti-tank unit soldier to be awarded the Knight's Cross. He was followed by *Leutnant* Pössinger and *Leutnant* Noak, who also distinguished themselves by repelling an enemy tank attack against the *1. Gebirgs Div.* and securing the Loire bridgehead near Beaugency.

During the French campaign, it was also found that the ammunition consumption in regimental companies was higher than in divisional anti-tank battalions, indicating that the latter were too immobile and not always able to be concentrated in time in the direction of the most critical threat.

The campaign in France clearly demonstrated that the 3.7cm *Pak* guns were no longer effective against increasingly well-armoured enemy tanks. As a result, immediate efforts were made to accelerate the development of new anti-tank gun designs, some of which were already in prototype form but not yet ready for mass production. In the meantime, the 3.7cm *Pak* would have to remain in service. To extend its battlefield relevance, a new armour-piercing round – the *Panzergranate 40*, (*Pz.Gr. 40*) – was introduced. Featuring a higher muzzle velocity and improved penetration, this tungsten-cored ammunition temporarily enhanced the gun's performance against better-armoured targets.

The Russian Campaign

On the eve of the war with Russia, in June 1941, the *Wehrmacht* possessed as many as 15,515 *Pak* guns. Despite the introduction of the new 5cm *Pak 38* from the end of 1940, the 3.7cm remained the primary German anti-tank gun. With the introduction of the new type of ammunition, *Pz.Gr. 40*, equipped with a tungsten core (the first 59,000 rounds were delivered in July 1940, and by July 1941, the stock of these rounds increased to 982,000), it was expected that the *Paks* would maintain their combat effectiveness.

This was indeed the case in the initial phase of the campaign, when older models comprised the overwhelming majority of Soviet armoured forces. During this period, the Red Army could field approximately 10,000 light tanks like the T-26 and around 7,500 fast BT tanks, alongside about 4,000 amphibious tanks such as the T-37, T-38 and T-40.

One notable example of the combat efficiency of the *Pak* in the early stage of the Russian campaign came about when *Gefreiter* (Lance-Corporal) Heinze from the *14. Kompanie* of *Inf.Rgt. 151* (part of *61. Inf.Div.*) destroyed ten Soviet T-26 tanks within just twenty minutes on 26 June near Windawa.

However, the *Pak* proved completely ineffective when facing the new types of Soviet medium tanks (T-34) and heavy tanks (KV-1/2), especially when these needed to be engaged from the front during their assaults.

The only really effective way to knock out those Russian machines was to damage their trucks with light guns, and then to get as close as possible to throw a large explosive charge on to the rear decking. These were desperate measures, and the losses in infantry, engineers and anti-tanks units who tried them were very high, but there was no other way until some better guns could be issued, and these were still some way off.

Soon, German soldiers themselves sarcastically dubbed the *Paks* as *Heeresanklopfgerät*, which can be translated as 'army door-knocking device', because the shells they fired simply bounced off the armour of Soviet tanks. In one report from June 1941 by the Fourth Army, it was noted that shells from the 3.7cm gun hit a T-34 tank twenty-three times without visible results.

Generaloberst Heinz Guderian, describing the dramatic battles near Moscow in the autumn of 1941 in his memoirs, recalled, among other things, the failure of the *112. Inf.Div.* Its units, defending against a Soviet tank counterattack on 17 November, were unable to stop it, leading to panic within the unit. One of the significant reasons he attributed to this was the fact that 'our 37mm anti-tank guns proved ineffective against the Russian T-34 tanks'.

Equally important, the Russians were fully aware of their tactical advantage. A Soviet report from 1942, describing tests of the penetrability of German guns, stated:

> 'The armour protection of the tank [T-34] at distances above 300 metres is sufficient against fire from German 37mm anti-tank guns. It should be noted that both *Pz.Gr.* and *Pz.Gr. 40* shells were used. From closer distances (200–300 metres), several penetrations of the hull sides and turret were achieved. Only one shell penetrated the front of the vehicle; it made a hole at the junction of the front hull plates.'

In this situation, it became clear that it was necessary to replace the 3.7cm *Pak* guns with larger-calibre anti-tank guns. Production of the 5cm *Pak 38* was already underway, and the 7.5cm *Pak 40* was soon to enter production.

Nevertheless, the 3.7cm *Pak* guns were still in use on the Eastern Front throughout 1942, albeit in decreasing numbers. An attempt to temporarily improve the situation was the introduction of the *Stielgranate 41* shaped-charge shells into service from February 1942, which were used both against Soviet armoured vehicles in 1943–44 and those of the western Allies during the fierce battles in Normandy in the summer of 1944, where the terrain favoured the use of short-range shells.

The new shell type turned the 3.7cm *Pak* into a grenade-launcher, with a projectile capable to penetrate even heavy tanks, but their use was tactically cumbersome. The *Stielgranate 41* was essentially a rod-type 'rifle grenade' loaded in the bore of the gun, with the finned sleeve sliding down outside the barrel.

The tail boom on this grenade was hollow and fitted over the muzzle of the gun, and the fins were attached to this boom. Inside the boom was a solid rod, which fitted into the bore and acted as the gas check for the propellant.

The bulbous nose of the bomb held 2.4kg of explosive, arranged as a hollow charge, and it could penetrate 150mm (6in) of armour. A type of blank cartridge was used to fire the bomb, and it must have been fairly inaccurate at any range greater than 180 metres (200 yards) because of the fact that it had virtually no barrel length to give it direction. Another drawback was that, to reload, one member of the crew had to go round to the front of the gun and slide another bomb on to the muzzle, so the chances of a second shot were just about nil. It would have taken uncommonly brave men to use it.

Starting from 1944, the *Pak* guns remained in use only with airborne and mountain divisions. Almost until the end of the war, they could still be found in various secondary units performing occupation or training duties.

Spanish Civil War, 1936-39

An anti-tank section of *Panzergruppe Drohne/Legion Condor* equipped with 3.7cm *Pak* guns, photographed during a march on the Vizcaya Front (Spain) in the summer of 1937.

(**Above**) Spanish Nationalist forces employed German-supplied *Pak* guns by organizing them into *baterías de cañón anticarro* (anti-tank batteries). These were typically attached at the battalion level, providing direct fire support against Soviet-supplied Republican armour, notably the T-26.

(**Opposite, above**) A Pak gun belonging to *Panzergruppe Drohne* being transported on the load bed of a truck, likely en route to a blacksmith's workshop for field repairs. This practice was common when guns required carriage or barrel maintenance, as moving them by truck reduced additional strain and ensured they could be repaired more quickly behind the lines.

(**Opposite, below**) Photo likely taken during the Battle of Alfambra in February 1938. Republican soldiers stand guard over a Soviet 45 mm M1932 (19-K) anti-tank gun – originally developed from the German 3.7cm *Pak* but enlarged in calibre to 45 mm – alongside two 37mm McClean Automatic Cannon Mk III, also supplied by the Soviet Union.

Soviet T-26 infantry tanks captured by the Spanish Nationalists were repurposed for use in their own tank battalion. These vehicles were armed with the 45 mm 20K gun, based on the German 3.7cm *Pak*.

The 20K, with an effective range nearly three times greater than the 3.7cm *Pak*, allowed Soviet crews to engage enemy armour from safer distances. Additionally, it boasted a higher muzzle velocity, which not only enhanced its armour penetration capabilities but also improved accuracy over longer ranges.

Poland, 1939

A 3.7cm *Pak* gun crew pictured during the early days of the German invasion of Poland in September 1939. The crew operates from a hastily prepared position, reflecting the mobile, fast-paced nature of *Blitzkrieg* tactics.

Withdrawing Polish infantry often set up ambushes along expected German advance routes, targeting the vulnerable towing vehicles. Such engagements frequently resulted in both combat and non-combat losses, and delays in movement.

By 1939, the *Wehrmacht* fielded four motorized infantry divisions, which were considered among the elite German formations. They were the only units in 1939 to be issued the *Sd.Kfz. 10* half-track vehicles to tow their 3.7cm *Pak* anti-tank guns, giving these divisions a tactical edge in the fast-paced, mechanized warfare of the early *Blitzkrieg* campaigns.

An anti-tank gun team from an *SS* regiment firing toward a Polish village. The presence of a soldier standing casually near a cottage suggests that the village was likely undefended.

Camouflaged in concealed positions, a *Pak* crew lays in wait for an anticipated advance of Polish forces. Although direct encounters with Polish armoured fighting vehicles were relatively rare, the anti-tank guns were still effectively employed against a range of softer targets such as trucks and staff cars, as well as lightly fortified defensive positions.

On 5 October 1939, a German victory parade was held in Warsaw to mark the successful conclusion of the campaign against Poland. The parade took place in front of Adolf Hitler, who made his only visit to the Polish capital that day.

(**Above**) Beginning on 9 September 1939, the Polish capital came under attack by the German *4. Pz.Div.*, which failed to capture the city in a direct assault from the march. Warsaw capitulated only on 28 September after heavy lighting. A *Pak* gun is seen positioned on the approach to Warsaw, in the city's Ochota district, on 9 or 10 September.

(**Opposite, above**) Lightly armoured Polish TK-3 tankettes were relatively easy to knock out, provided they were spotted, which was not always easy due to their low silhouette. This photo was taken on the outskirts of Warsaw in mid-September 1939.

(**Opposite, below**) A Polish twin-turret 7TP tank, damaged and abandoned following an advance on Okęcie, which was repelled by German *Pak* guns on 12 September 1939.

France and the Low Countries, 1940

During the invasion of the Low Countries, *Pak* guns were ferried across rivers and canals using small rafts placed on nimble dinghies. This method helped sustain the speed of the German offensive and enabled the *Pak* teams to provide vital support to advancing infantry units.

Soldiers of a *Pz.Jgd.Abt.* from the *264. Inf.Div.* are caught in a traffic jam while attempting to cross a burning village in the Netherlands.

A 3.7cm *Pak* deployed in a firing position within a French town, placed at a crossroads to maximize its field of fire. The strategic siting allowed the crew to cover multiple approach routes simultaneously, making it an effective ambush point against enemy armour or motorized columns.

Traffic congestion during the French campaign of 1940, with a mix of tracked vehicles, cars and towed *Pak* guns jammed together on a narrow road. Such scenes were common as German spearheads pushed deep into France, with rapid advances often creating bottlenecks that temporarily slowed entire columns.

A *Pak* gun crew takes a short combat break, appearing relaxed and engaged in casual conversation. Such breaks were rare and valuable, offering the men a chance to rest, smoke and exchange thoughts before being ordered back into action.

A *Pak* section of the *9. Pz.Div.*, using an Austrian-built *Steyr 640* truck as a gun tractor and cross-country lorry, photographed in Saint-Just in May 1940.

A German column advancing towards Rosières-en-Santerre, 110km north of Paris, where an RAF airfield was built for the 18th and 57th Bomber Squadrons. When the Battle of France was over, the Rosières airfield became home to *III./KG1*, a Luftwaffe unit flying the Heinkel He-111H bomber.

A *Pak* gun protecting the bridge over the Meurthe Canal in Nancy, which was captured by the Germans during the early stages of the French campaign. From dusk to dawn, only half of the crew stood guard at the gun, while the rest slept nearby.

A *Pz.Abw.Kp.* passing by the ruins of an unidentified French town. While the tanks spearheaded the German advances, it was the field artillery and air force that inflicted the heaviest damage on infrastructure.

A 3.7cm *Pak* from *Panzergruppe Guderian* (note the large 'G') in Mulhouse, a city in Alsace, eastern France, near the German border.

The *Pak* gun was purposely designed to allow its crew to move it around the battlefield without a towing vehicle. This photo was taken in May 1940 close to Dunkirk, France, by Hermann Weper, an officer assigned to *MG Abt. 52*.

On 24 May 1940, Hitler issued the famous 'Halt Order', temporarily stopping the advance of German panzer divisions outside Dunkirk. This pause allowed the bulk of the British Expeditionary Force to organize its defence and ultimately evacuate across the Channel. During this standstill, German *Pak* crews dug in their guns, prepared for possible counterattacks and waited for the next move.

(**Above**) During the Battle of France, 3.7cm *Pak* crews often found their rounds bouncing harmlessly off French Char B1 tanks, a failure that earned the gun its infamous nickname *Heeresanklopfgerät* ('army door-knocking device').

(**Opposite, above**) The *Pak* was also nearly useless against the French Somua S-35 tank, which boasted excellent armour protection (up to 47mm thick), unless their crews could land a shot on the rear or side at close range, or disable the tracks.

(**Opposite, below**) The British Matilda II, known officially as the Infantry Tank Mk II, had frontal armour up to 78mm thick, making it virtually impervious to all standard German anti-tank weapons in 1940.

Preparing for the invasion of Britain, 1940

(**Above**) A German *Pak* crew manning a firing position hidden within the confines of a simple French beach cottage, its barrel pointing unwaveringly across the Channel toward Britain.

(**Opposite, above**) On the shore, engineers laboured over improvised invasion rafts, built upon massive pontoons lashed together. It looked crude, yet it carried grim purpose: designed to be towed across the Channel behind landing craft and barges, ferrying men and guns into the storm of fire awaiting them on Britain's beaches.

(**Opposite, below**) A barge armed with a 3.7cm *Pak* on an improvised basis during preparations for Operation Sea Lion, the planned German invasion of Britain.

(**Above**) A German *Pak* gun crew is pictured in a stark desert environment, where the earth is dry and barren, covered not with green grass but with brittle, sun-scorched tufts of yellow.

(**Opposite, above**) A moment of tense anticipation – the calm before the storm. An *Sd.Kfz. 250* half-track of the headquarters section, *Schützen-Rgt. 104, 21. Pz.Div.*, fitted with additional stowage boxes and equipped with extended radio equipment, waits in silence for the enemy's approach.

(**Opposite, below**) In the chill of a desert dawn, a *Pak* crew is ready for action. Though the Sahara is synonymous with blistering heat, the early morning brings biting cold, with temperatures plunging to near 0°C.

Soviet Union, 1941-1944

The capture of Grodno by German forces in June 1941. During the first day of Operation Barbarossa, the city was taken from the Soviets, who had occupied it since September 1939, following the attack on Poland. (NAC)

While advancing through Belarus, a region dominated by dense forests and expansive marshlands, German military columns encountered significant logistical challenges. The terrain offered very limited infrastructure, with only a handful of narrow, poorly maintained roads available for movement. The PaK gun shown bears the markings of *Pz.Jg.Abt. 17*, the anti-tank battalion of the *17. Inf.Div.*

Prepared defensive positions for *Pak* guns were something of a luxury when a fast-moving battle was being fought. In the East, anti-tank guns were often positioned in roadside ditches, which offered only limited protection.

A German *Pak* gun covering the advance of a captured French Char B1, redesignated as *PzKpfw. B2 740(f)*, during operations against the Soviet 'Stalin Line' in the summer of 1941. Although the B1's low speed made it unsuitable for rapid armoured thrusts, the 75mm hull-mounted gun remained highly effective for engaging Soviet bunkers and fortified positions.

A *Pz.Jgd.Kp.* from an unidentified *SS* division enters a Russian village in the autumn of 1941. While most anti-tank units were equipped with wheeled gun tractors, *SS* formations were provided with light tracked tractors (*SdKfz. 10*), which proved to be much more versatile on the rough, dirt, and muddy roads of the Eastern Front.

An anti-tank company of the *169. Inf.Div.* advancing on the Finnish Front, summer 1941. On 8 Juy, the division captured the town of Salla. In coordination with the Finnish 6th Division, the *169. Inf.Div.* subsequently pushed Soviet forces back beyond the pre-Winter War frontier. The *169. Inf.Div.* was raised in November 1939 as a unit of the 5th Mobilization Wave, with its anti-tank companies initially horse-drawn, but by the summer of 1941 they had been partially motorized with requisitioned civilian trucks to improve mobility in the challenging northern terrain.

During the harsh Soviet winter, *Pak* crews had to contend with the brutal and freezing cold. Greatcoats and balaclavas, worn under helmets, were used to mitigate cold, but were often not enough to withstand the extreme temperatures. Animal skin coats (often referred to as *pelzmäntel* or 'furs'") became a sought-after commodity

The winter conditions of 1941/42 significantly complicated the task of setting up anti-tank gun positions. One key method was to cover the *Pak* protection shields with whitewash, made from a mix of lime and water, a quick and inexpensive solution that helped blend the guns into the snowy landscape.

The winter often immobilized motorized towing vehicles, forcing even staff cars into improvised use as gun tractors. With the arrival of deep snow, however, horses and ski-mounted chassis frequently proved to be the most reliable means of moving light anti-tank guns across difficult terrain.

With the onset of the spring of 1942, the obsolete yet lightweight *Pak* guns proved to be the only artillery that could be relatively easily towed across the difficult terrain.

A German *Pak* crew fires during the 1942 advance on Sevastopol. The lack of fieldworks left the gun team highly vulnerable to enemy counter fire, but the urgency of the assault on the Crimean fortress required crews to fight from exposed positions to provide immediate anti-tank and fire support.

A *Pak* crew photographed somewhere on the northern coast of the Black Sea during the fighting at the Kuban which took place from February–July 1943. The battle was part of the larger strategic context of the German defeat at Stalingrad and the subsequent Soviet counteroffensive.

A *Pak* gun on the Eastern Front being hauled on a sleigh by a pair of horses in January 1942. The brutal winter conditions rendered many motorized vehicles useless, forcing German units to rely on improvised transport methods. (NAC)

During the second winter of operations on the Eastern Front (1942/43), some anti-tank units were already equipped with the *Sd.Kfz. 2 Kettenkraftrad*, a compact, tracked, cross-country vehicle employed to tow light artillery pieces mounted on improvised skis. They proved highly effective in harsh winter conditions. By 1945, approximately 9,000 *Kettenkräder* had been produced, serving primarily in the transport of light loads, including machine guns, light artillery, ammunition and essential supplies.

A *Pak* concealed in deep snow, with its crew wearing white camouflage over their uniforms. The gun's low silhouette made it almost invisible against the winter landscape, offering a hidden yet deadly threat to advancing enemy light armour.

With helmets set aside and a horse and cart passing peacefully in the background, the scene suggests a lull in the fighting and little expectation of an immediate enemy counterattack. Such static guard positions were common in rear or flank areas, where anti-tank crews doubled as local security against sudden armoured thrusts.

Soviet forces, familiar with the local terrain, often launched counterattacks through marshes and over rivers, sometimes transporting their light artillery on makeshift rafts. German *Pak* crews proved highly effective in these situations, as their guns could target the slow-moving, exposed Soviet craft on the water.

Captured *Pak* guns after the Battle of Kursk in the summer of 1943. The weapons still bear tactical markings introduced in the mid-1930s, such as the letter code identifying their place within the company (e.g., 'H' for the fourth gun in the 2nd Platoon). In the background, a Soviet ZiS-5 truck tows a 61-K 37 mm anti-aircraft gun.

Chapter Seven

Panzer III on the Front Lines

When the German Army entered the Second World War in September 1939, the *Panzerwaffe*'s armoured fist was still a work in progress. Contrary to the later image of German tanks dominating through sheer firepower, the majority of vehicles in panzer divisions at the outbreak of the war were light tanks – the *PzKpfw. I* armed only with machine guns and *PzKpfw. II* equipped with a 2cm autocannon. The true anti-tank punch was entrusted to a relatively small number of the *PzKpfw. III*, whose main armament was the 3.7cm *KwK*, adequate against most European tanks, whose armour rarely exceeded 30mm.

Mounted in a fully traversable turret, the *KwK L/45* gave the *PzKpfw. III* the ability to act as the designated tank-killer. In theory, Panzer III units would lead armoured spearheads, engaging enemy tanks head-on while lighter panzers flanked and exploited breakthroughs. The gun's relatively high muzzle velocity allowed for accurate gunnery, while its compact size made it ideal for the *PzKpfw. III*'s three-man turret, a design feature that maximized rate of fire (twenty per minute) and crew efficiency.

In practice, however, *PzKpfw. III* tanks made up only about 3 per cent of the roughly 2,700 armoured fighting vehicles deployed by German forces against Poland. Their operational role in the campaign was therefore limited, but since the Polish Army's armoured force consisted of only 134 7TP light tanks armed with the 37mm Bofors gun, the *PzKpfw. III* still enjoyed a notable qualitative advantage. This was further reinforced by the *Panzerwaffe*'s deployment of nearly 170 ex-Czech tanks – the *PzKpfw. 35(t)* and *PzKpfw. 38(t)* – which were also armed with 37mm guns and provided substantial additional firepower.

The Polish Campaign, 1939

On 1 September 1939, the *Wehrmacht* fielded ninety-eight *PzKpfw. III* tanks, of which thirty-eight were of the new *Ausf. E* version. Prior to that date, sixty tanks from the earlier production series had been completed: ten *Ausf.* As, ten *Ausf.* Bs, fifteen *Ausf.* Cs and twenty-five *Ausf.* Ds. This total does not include the retroactively assembled hybrid *PzKpfw. III Ausf. D/B*, which was delivered only in the autumn of 1940.

Of the ninety-eight *PzKpfw. III*s in service on 1 September 1939, the largest concentration – thirty-seven vehicles – was assigned to I Battalion (*Abt.I*), *Panzer-Lehr-Regiment* (*Pz. Lehr Rgt.*), which primarily operated the *Ausf.* E. Another eleven tanks were held by the two training units mentioned earlier. *Pz.Rgt. 1*, based in Erfurt, and part of the *1. Pz.Div.* fielded twenty tanks, including early *Ausf. A* and *B* variants. During the Polish campaign, the division served with XVI Army Corps, Tenth Army, Army Group South, advancing along the main axis toward Warsaw and engaging in heavy fighting. Later, it participated in the encirclement of Polish forces in the Radom area.

Pz.Rgt. 2, based in Ohrdruf, Thuringia, with six *PzKpfw. III Ausf. C–D*, also belonged to the *1. Pz.Div.* At this stage of the war, panzer divisions still comprised two tank regiments, so *Pz.Rgt. 2* followed the same combat route as *Pz.Rgt. 1* throughout the campaign.

All other panzer regiments had only three *PzKpfw. III*s each, including at most one *Ausf. E*; the remainder were generally *Ausf. C* and *D*, except for *Pz.Rgt. 5*, which operated *Ausf. A* and *B* models. *Pz.Rgt. 5* was among the first units to receive the *PzKpfw. III* – together with *Pz.Rgt. 1* – and was thus one of the earliest equipped. Stationed in Wünsdorf near Berlin, *Pz.Rgt. 5* formed part of the *3. Pz.Div.*, alongside *Pz.Rgt. 6* from Zossen. Both regiments had three *PzKpfw. III*s and nine *PzKpfw. IV*s each, plus sixty-three and fifty-nine *PzKpfw. I*s, and seventy-seven and seventy-nine *PzKpfw. II*s, respectively – illustrating the limited numerical role of the *PzKpfw. III* in the 1939 campaign.

The *3. Pz.Div.*, with its *Pz.Rgt. 5* and *6*, was assigned to XIX Panzer Corps, Fourth Army, Army Group North. It advanced from central Pomerania into the area north of Bydgoszcz, then moved south of Białystok before turning toward Brest from the north. The division bypassed Brest and continued toward Włodawa.

Pz.Rgt. 3 and *4*, stationed in Dresden and Wien-Laxenburg/Wiener Neustadt respectively, each possessed three *PzKpfw. III*s, a large number of *PzKpfw. I* and *II* models, and eight or nine *PzKpfw. IV*s. These units formed the *2. Pz.Div.*, assigned to XVIII Army Corps, Fourteenth Army, Army Group South. The division advanced toward Lwów from the southwest, was later transferred to XVII Army Corps, and turned northwest to fight in the Rawa Ruska–Tomaszów Lubelski–Zamość area. In the final stage of the campaign, it reached the Szczebrzeszyn area northeast of Sandomierz.

Pz.Rgt. 7 from Eisenach, with three *PzKpfw. III*s, served in the improvised *Pz.Div. 'Kempf'*, a mixed formation of *Wehrmacht* and *SS* units. In August 1939, the division was assigned to I Army Corps, Third Army, Army Group North, in East Prussia. It entered combat at Mława, suffering heavy losses – seventy-two of its 164 tanks, the highest of any German panzer division in Poland. Thereafter, it advanced toward Ciechanów and Różan, and in the Maciejowice area took part in the encirclement of Polish forces retreating toward Warsaw. From 19 September, the division participated in the siege of Modlin. After the campaign, the 7th Regiment was reassigned to the newly formed *10. Pz.Div.*

Pz.Rgt. 8 from Böblingen, also with three *PzKpfw. III*s, belonged to the *10. Pz.Div.* After transferring *Pz.Rgt. 7* to *Division 'Kempf'*, the division retained only one tank regiment. Initially in Fourth Army reserve, it crossed the border on 3 September, moved through Gdańsk into East Prussia without opposition, and first engaged in combat on 7 September near Łomża. The next day it helped break Polish defences at Wizna, then fought near Zambrów and Bielsk Podlaski before advancing toward Brest from the northwest. After the capture of Brest, the division's combat in Poland came to an end.

From *Pz.Rgt. 10* at Zinten (now Korniewo, Kaliningrad Oblast), only I Battalion participated in the campaign, bringing three *PzKpfw. III*s. It operated as a reserve unit under XXI Army Corps, initially supporting the *21. Inf.Div.* in fighting against the Polish 16th Infantry Division along the Ossa River.

Pz.Rgt. 15 from Sagan (now Żagań), part of the *5. Pz.Div.* from Opole, also fielded three *PzKpfw. III*s. This division, serving in VIII Army Corps, Fourteenth Army, Army Group South, entered Poland near Rybnik in Silesia and advanced east toward the Vistula near Sandomierz. In mid-September, it moved toward Stryj, where it made contact with Soviet forces advancing from the east.

The I Battalion of *Pz. Lehr Rgt. 'Wünsdorf'* also took part in the campaign. In December 1938, the regiment transferred its fully equipped I Battalion to *Pz.Rgt. 33*, while its incomplete second battalion was disbanded. During the 1939 campaign, I Battalion of *Pz. Lehr Rgt.* 'Wünsdorf' was attached to the *3. Pz.Div.*, XIX Panzer Corps, Fourth Army, sharing its combat route.

In total, the Germans lost thirty *PzKpfw. III* tanks during the Polish campaign, four of which were later repaired, giving a net loss of twenty-six. Since Fourth Army had suffered no *PzKpfw. III* losses by 25 September, it can be assumed that nearly all of these were *Ausf. A–D* models. The heaviest losses – eleven tanks – were sustained by the *1. Pz.Div.*, which was engaged along the Tenth Army's main axis of advance.

The losses among the *PzKpfw. III* tanks were surprisingly high, considering that out of ninety-eight engaged in combat, twenty-six were lost. Extensive after-action reports, written by officers from armoured regiments at all levels, confirmed that while the *PzKpfw. III* was combat-effective, it required stronger armour.

Just over thirty surviving *PzKpfw. III Ausf. A–Ds* from the Polish campaign accompanied their parent units westward to face France. However, between February and March 1940, nearly all were replaced by the newer *PzKpfw. III Ausf.* E, built in a series of ninety-six units (serial numbers 60401–60496). Although still armed with the 3.7cm gun, the *Ausf. E* featured significantly thicker front armour of 30mm. The withdrawn *Ausf. A–D* models were reassigned to training units in Germany, where they remained in service until they were completely worn out.

The *Ausf. E* was followed by the *Ausf. F* (3.7cm gun, serial numbers 61001–61636) and the *Ausf. G* (3.7cm gun, serial numbers 65001–65950). By the end of April 1940, at the outbreak of the campaign in the West, the total number of *PzKpfw. III Ausf. E* and *G*s amounted to approximately 380 vehicles.

None of the *PzKpfw. III* tanks participated in the Norwegian campaign (the occupation of Norway and Denmark) between April and June 1940.

France and the Low Countries, 1940

The main offensive against France was carried out by Army Group A under *Generaloberst* Gerd von Rundstedt, which included the Fourth Army on the right, northern flank, and the Twelfth Army on the left, southern flank.

The Fourth Army comprised the XV Motorized Corps, which included the *5. Pz.Div.* with twenty-four *PzKpfw. III* tanks in *Pz.Rgt. 15* and a further twenty-eight in *Pz.Rgt. 31*; the *7. Pz.Div.* had no *PzKpfw. III*s.

Panzergruppe Kleist (Twelfth Army) contained the XIX Motorized Corps (under Guderian), made up of the *1. Pz.Div.* (twenty-eight *PzKpfw. III*s in *Pz.Rgt. 1* and thirty in *Pz.Rgt. 2*), the 2. Pz.Div. (twenty-nine *PzKpfw. III*s in *Pz.Rgt. 3* and twenty-nine in *Pz.Rgt. 4*) and the *10. Pz.Div.* (twenty-nine *PzKpfw. III*s in *Pz.Rgt. 7* and twenty-nine in *Pz.Rgt. 8*). Also within the Twelfth Army was the XXXXI Motorized Corps (General of Armoured Troops Georg-Hans Reinhardt), which included the *6.* and *8. Pz.Div.*, both without any *PzKpfw. III* tanks.

On the secondary northern front, Army Group B under Generaloberst Fedor von Bock operated, consisting of the Sixth Army, advancing through central Belgium, and the Eighteenth Army, advancing through the Netherlands. The Sixth Army included the XVI Motorized Corps, with the 3. Pz.Div. (forty-two PzKpfw. IIIs in Pz.Rgts 5 and 6) and the 4. Pz.Div. (twenty PzKpfw. IIIs in Pz.Rgt. 35 and twenty in Pz.Rgt. 36). The Eighteenth Army included only the 9. Pz.Div., which fielded forty-one PzKpfw. III tanks in its single Pz.Rgt. 33.

During the crossing of the Meuse near Sedan, the counterattack at Arras and other intense battles of the campaign, German forces swiftly defeated the Allies, culminating in the armistice signed on 22 June 1940 (which came into effect two days later). German panzer divisions participated in both phases of the campaign—the initial *Fall Gelb* ('Case Yellow'), after which they were withdrawn from the front on 28 May for rest and refitting, and the second phase, *Fall Rot* ('Case Red'), from 5 June until the armistice.

The *PzKpfw. III Ausf. F* and *G* proved highly vulnerable to enemy anti-tank fire. The French 2.5cm anti-tank gun could penetrate the armour of a *PzKpfw. III* from a distance of 500 metres, while the 4.7cm anti-tank gun was capable of penetrating its armor even at 1,500 metres. As a result, during the French campaign, the Germans lost 110 *PzKpfw. III* tanks in May and an additional twenty-five in June, for a total of 135 losses. Only a portion of these losses were repaired; over a hundred were lost permanently.

The 3.7cm gun also proved problematic. On 4 June 1940, the *Inspekteur der Panzertruppen 6* reported that combat experience in the West had shown the 3.7cm *KwK* mounted on the *PzKpfw. III* was inadequate as an armour-piercing weapon. Consequently, the head of the *OKH* (*Oberkommando des Heeres*, the Army High Command) requested the accelerated introduction of the 5cm *KwK* in *PzKpfw. III* production.

The very next day, Krupp's *Abteilung A.K.* completed drawings for mounting the 5cm *KwK* in the *PzKpfw. III* turret, and by the end of June, the first five *Ausf. G* tanks equipped with the 5cm gun were produced. Until October 1940, both 3.7cm and 5cm-armed tanks were produced, but thereafter, all *PzKpfw. III* tanks were outfitted exclusively with the 5cm *KwK*.

Nevertheless, by December 1940, 483 *PzKpfw. III* tanks armed with the 3.7cm gun remained in front-line units. With no immediate replacements available, these tanks had to be deployed in upcoming campaigns despite their shortcomings.

The Balkan Campaign, 1941

On 6 April 1941, German forces launched *Unternehmen Marita* (Operation Marita), crossing first into Yugoslavia and then Greece. Six panzer divisions took part in the invasion, fielding approximately 850 tanks, of which 109 were *PzKpfw. III*s armed with the 3.7cm *KwK*. The largest number of these tanks (twenty-seven) was with the *2. Pz.Div.*, while the newly formed *14. Pz.Div.* had the fewest with sixteen.

The campaign was brief and largely successful. Only eleven days after the invasion began, the Yugoslav government in Belgrade capitulated. Greece was occupied soon afterwards. British and Commonwealth forces, unable to hold their positions in Greece, withdrew and were then evacuated from Crete to Egypt on 1 June, abandoning all their heavy equipment.

Much of the fighting took place in rugged, mountainous terrain strewn with rocks, which placed severe strain on the tanks' mechanical systems. Failures of the final drive units, drive sprockets and leaf springs were frequent. The most serious issue, however, was the rapid disintegration of the rubber tyres on the running wheels, which wore out after only a short period of use. Despite these problems, the campaign demonstrated that tanks could still be effectively employed even in such inhospitable conditions.

The Invasion of the Soviet Union, 1941–1942

For the invasion of the Soviet Union – codenamed Barbarossa – which began on 22 June 1941, the German Army assembled roughly 3 million men and seventeen panzer divisions, equipped with over 3,200 tanks, including 259 *PzKpfw. III*s armed with 3.7cm guns and 707 with 5cm guns. These forces faced a Soviet armoured force three to four times larger, which included the formidable KV heavy tank and the revolutionary T-34 medium tank, both of which were immune to the 3.7cm gun, with superior armour protection and more powerful main guns.

Fortunately for the Germans, in the first weeks of the campaign the element of surprise crippled large parts of the Red Army, which had been weakened by Stalin's purges, lacked coherent strategy and showed poor tactical employment of its armour.

However, as Soviet defences steadily stiffened, losses of 3.7cm-armed *PzKpfw. III* tanks rose sharply. Out of 350 of these in both front-line service and reserve in June 1941, only 203 remained operational by the end of 1941. By the spring of 1942, all remaining vehicles of this type had been withdrawn from combat and replaced with *PzKpfw. III*s mounting 5cm guns or with the *PzKpfw. IV*. The last surviving 3.7cm-armed *PzKpfw. III* tanks were relegated to training units in Germany, where they remained until completely worn out.

It is worth noting that during the first month of Operation Barbarossa, five older hybrid *PzKpfw. III Ausf. D/B* tanks – completed in October 1940 – were issued to the independent *Heerestruppen* unit *Panzer-*

Abteilung z.b.V. 40. These were assigned to the XXXVI. Armeekorps under *A.O.K. Norwegen* (Army Command Norway), where they supported both German and Finnish forces in operations against the Soviets. After sea and rail transport to the front, they fought in the Kuusamo–Sohjana–Kiestinki and Salla–Kandalaksha sectors, aiding Finnish advances and securing key points such as Vuokkiniemi and Oinasniemi. Harsh terrain, mines, weak bridges and mechanical breakdowns limited their mobility, and by early 1942 most required major repairs. A shortage of spare parts kept them from full operational readiness. In summer 1942, reinforced with new *PzKpfw. III* tanks, the unit took part in the Finnish–German Ahvenlahti offensive, but deep snow and Soviet defences stalled progress. By late 1942, the battalion was ordered to Norway, leaving a company – *Panzer-Kompanie 40* – in Finland with a handful of *PzKpfw. III* (likely both older *D/B* and *E* models) and *PzKpfw. I* tanks, continuing limited operations under the XVIII Mountain Corps into 1943.

Czechoslovakia, 1938-39

The annexation of the Sudetenland was carried out peacefully, with crowds of ethnic Germans welcoming German tanks with flowers – earning the campaign the nickname 'The Flower War'. The tank in the photo likely belongs to the *Kp.5/2.Pz.Div.*

During the annexation of the Sudetenland in autumn 1938, a *PzKpfw. III Ausf. D* passes through a ceremonial gate built by ethnic Germans, decorated with the slogan 'One Nation, One Country, One Leader'. The first tank ('R01') belongs to the regimental commander, while the second ('R02') is assigned to the radio officer, serving as a backup vehicle for the commander. If the CO had to take over 'R02', the radio officer would be reassigned to another vehicle.

The annexation of Moravia in mid-March 1939 took place under exceptionally difficult weather conditions, with heavy snowfall and icy roads. Nevertheless, upon entering German-populated areas, panzer troops (*3. Pz.Div.*) were warmly welcomed by the local ethnic German population.

(**Above and below**) A *PzKpfw. III* followed by *PzKpfw. IIs* of the *1. Pz.Div.* pass by *Generalleutnant* Geyr von Schweppenburg during a military parade in Prague's Wenceslas Square on 17 March 1939.

The parade in Wenceslas Square followed the swift invasion of Bohemia and Moravia, which had been transformed into a German protectorate. For half an hour, German tanks of all classes rolled down the square, while the pleasant weather drew large crowds to watch from both sides. Note the photo in the middle, which depicts a *PzKpfw. III Ausf. D* with the tactical number 'R01', identifying it as the regimental commander's tank.

Poland, 1939

(**Opposite, above**) A *PzKpfw. III Ausf. E* of the *Pz. Lehr Abt.* featuring unusual front armor markings (Nazi eagle). This tank advanced into central Poland from East Prussia in September 1939 with the *3. Pz.Div.*

(**Opposite, below**) The tank commander occupied the rear of the turret, seated centrally between the gunner and loader, with the quickest access to the top hatch for observation or escape. His subordinates, by contrast, made use of the smaller side hatches of the turret. Note that by September 1939, some *PzKpfw. III*s were retrofitted with an *N.K.A.V.* smoke grenade rack mounted on the rear of the hull.

A *PzKpfw. III* of *Pz.Rgt. 15, 5. Pz.Div.*, during the Polish campaign. The tank belonged to *Abteilung I*, which can be identified by the colour of the rhomboid painted beneath the turret tactical number ('13') – red for the 2nd Company or, more likely in this case, blue for the 4th Company.

A *PzKpfw. III Ausf. D* in service with the brigade staff of the *4. Pz.Div.* The frame antenna for the long-range radio is mounted on non-conductive wooden rods. This radio tank carried only an *MG 34* in a ball mount in the turret; all other weapons were dummies. The Panzer III is accompanied by a *Befehlspanzerkampfwagen 265*, based on the *PzKpfw. I* chassis.

A *PzKpfw. III Ausf. E* of the *Pz.Lehr Abt.* with tactical numbers painted on the hull sides and rear. This vehicle was photographed while shelling a village defended by Polish infantry near the Polish–East Prussian border.

A surprising duel between *PzKpfw. III Ausf. Es* from the *Pz.Lehr Abt.* and a Polish PZL P.11c fighter from the 151st Fighter Squadron, which was forced to land near Zambrów, close to the East Prussian border.

France, 1940

A *PzKpfw. III*, stuck in a traffic jam during the advance in France, leads a column of lightly armed and armoured *PzKpfw. II*s. This photograph is unusual, as during the French campaign German foot soldiers rarely kept pace with the rapid advances of their armoured units. Tanks often operated with minimal infantry support, relying on speed and surprise rather than close coordination with infantry.

A rare moment of tank crew respite, soon to end as a motorcycle messenger arrives with new orders.

Top view of a *PzKpfw. III* turret and open engine hatches. The tank is undergoing a quick post-combat check before being sent back into battle.

A rather surprising mix of *PzKpfw. III*s and a *PzKpfw. I*, the latter already obsolete and of minimal combat value by the French campaign. The Panzer I in the photograph is even disarmed. Taken in a maintenance area, the image highlights how these light tanks were often repurposed as non-combat vehicles, such as transporting spare parts or towing damaged machines, rather than participating directly in front-line action.

An interesting photograph capturing field maintenance of a *PzKpfw. III* from the *3. Pz.Div.*, which required removing the entire turret, including the 3.7cm *Pak* gun. Maintenance companies were able to perform this task thanks to mobile cranes capable of lifting loads of several tons.

A brief combat pause for refuelling using jerry cans. During movement, panzer divisions rarely had the opportunity for resupply from mobile fuel trucks, so all tank types had to carry additional fuel with them. This practice allowed armoured units to maintain their rapid advance over long distances, but also added extra weight and exposed tanks to greater vulnerability.

Balkans, 1941

A *PzKpfw. III* from the *9. Pz.Div.*, part of Army Group C under *Generaloberst* Wilhelm List, during the Balkan campaign of April 1941. Note that, while entering Yugoslavia, many German tanks carried spare tracks and wheels as provisional additional frontal armour.

A *PzKpfw. III* of the *11. Pz.Div.*, with the *Gespenster* ('Ghost') symbol, during a brief combat pause, likely in a Greek village, in April 1941. Tanks were less decisive in the outcome of this part of the campaign due to the mountainous terrain and poor roads, but motorcycle units proved highly effective; fast and manoeuvrable, they allowed for rapid advances.

Although Greece had no armoured force capable of opposing German panzer units, some *PzKpfw. III*s were lost, mostly to anti-tank guns. The tank in the photo was knocked out somewhere in the Thermopylae region.

Soviet Union, 1941–1942

(**Opposite, above**) To manage the long distances of marches across the Soviet Union, *PzKpfw. III*s were often fitted with jerry cans of fuel on the rear of the hull and atop the turret, which made them vulnerable to even small-calibre weapons.

(**Opposite, below**) During the initial stage of Operation Barbarossa, fast-moving *PzKpfw. III*s often advanced ahead of the infantry, with soldiers riding on the hull to help protect against enemy ambushes and engage in close combat if necessary. By August 1941, no solid defensive line existed on many sections of the front, so the Germans aimed to advance as far as possible, encountering only light opposition.

With the approach of the first Soviet winter, which proved to be the coldest of the twentieth century, German panzer troops were issued with winter coats to wear over their black combat uniforms. By December, however, it had already become clear that these garments would not provide sufficient protection against the extreme cold.

During the autumn of 1941, all German tanks were frequently immobilized by the heavy mud of the Eastern Front, a period known as the *Rasputitsa*. *PzKpfw. III*s, weighing around 20 tons, were particularly affected due to their size and weight. Crewmen often had to perform labour-intensive cleaning of the chassis and tracks to restore mobility. The most efficient method involved the use of fire-fighting water carts, where available, to wash away the thick layers of mud and debris clogging the suspension and running gear.

(**Above and below**) *PzKpfw. III Ausf. D/B*s (hybrid) from *Kompanie 3, Panzer-Abteilung 'Zum Besonderen Verwendung'*, in Kiestinki, near the Soviet border with Finland, during May 1942. The unit was deployed in northern Finland as part of Germany's efforts to secure key positions along the northern front. Operating in harsh terrain, it carried out infantry support, often adapting its vehicles to cope with limited infrastructure and challenging conditions. (SA-kuva)

Chapter Eight

Beyond Wehrmacht

The 3.7cm calibre guns produced by Rheinmetall were exported both before the outbreak of war and during wartime operations.

Soviet Union

The design of anti-tank guns in the USSR began in the late 1920s, but due to a lack of experience in developing such artillery systems, Moscow sought a foreign partner that would offer technical assistance. Meanwhile, Germany needed opportunities to continue secret work on promising designs and maintain its scientific and technical potential despite the Versailles Treaty restrictions.

In 1929, Rheinmetall created a front company for contacts with the USSR called 'BUTAST' (Bureau for Technical Works and Studies Ltd/ *Bureau für technische Arbeiten und Studien*). The main task of this company was to design new artillery systems and test them outside Germany to avoid detection by the Allied disarmament control commissions.

On 6 August 1930, the Soviet of People's Commissars passed a resolution, and a secret agreement was signed in Berlin on 28 August. Under this deal, the Germans committed to helping the USSR organize mass production of six artillery systems, ranging in calibre from 37mm to 152mm. Rheinmetall was to deliver prototypes of the anti-tank guns, complete sets of documentation and kits of semi-finished products for assembling several units of each type in the USSR. The agreement included clauses stipulating that the USSR must not disclose the deal's terms or the specifications of the German guns, and Germany must not reveal any information it learned about Soviet artillery factories. Rheinmetall was paid $1.125 million for these services.

The German side fully met the terms of the deal regarding the 3.7cm gun, delivering twelve *Tak* guns to the USSR. The Russian copy of *Tak*, designated as 1-K, went into mass production at the Kalinin Plant No. 8 in Podlipki, near Moscow.

The factory faced enormous difficulties in mastering the production, with weapons being made in a semi-handcrafted manner with manual fitting of parts. In 1931, the factory presented 255 guns to the Red Army, but none were accepted due to poor assembly quality. In 1932, they managed to deliver 404 guns, and in 1933 another 105. Production of these guns was discontinued in 1932 (the guns delivered in 1933 were those not accepted the previous year) due to the adoption of the more powerful 45mm anti-tank gun Model 1932 (19-K), which was a development of the 1-K.

Based on the 1-K, the 37mm B-3 (5-K) tank gun was created, which armed the BT-2 and T-26 tanks.

As of 1 January 1936, the Red Army had 506 1-K guns, comprising 422 in good condition, fifty-three requiring major repairs, twenty-eight training guns and three that were unserviceable. It is likely that in subsequent years, the number of serviceable guns decreased, and they were gradually moved to storage and training institutions. Information on their combat use before 1941 is not available.

The exact number of 1-K guns in Soviet first-line formations as of 1 June 1941 is unknown, but it is certain that some were present in units such as the VIII Mechanized Corps. At the start of the war, all guns in storage were likely thrown into combat due to a shortage of artillery to equip newly formed units and replace enormous losses. No specific details of their combat use have been established, and it is likely that almost all were lost in 1941. According to some reports, there was a shortage of shells for this type of gun.

The German military captured a number of Soviet 1-K anti-tank guns and integrated them into their own arsenal. These guns were re-designated by the Germans as the 3.7cm *Pak-158(r)* ('r' standing for *russisch*, or 'Russian').

In 1937, the 1-K chassis was used as the basis to mount an improved version of the 19-K gun. The 19-K itself had been the Soviet adaptation of the German 3.7cm *Pak*, enlarged to 45mm calibre. The new design incorporated several important improvements: a semi-automatic breech, a modernized sighting system, a firing button mechanism, better suspension, a more reliable shield mount and adjustments to the balance of the moving parts.

The sum of these evolutionary changes resulted in a significantly different weapon, enough to warrant a new factory designation: 53-K. After minor defects were corrected during testing, the gun was cleared for service and formally adopted on 24 April 1938 as the 45mm M1937 anti-tank gun. Production at Plant No. 235 continued uninterrupted until 1944.

In organizational terms, the 53-K was widely distributed throughout the Red Army. Two guns formed the standard anti-tank platoon assigned organically to every rifle battalion. At the divisional level, each rifle division fielded a full anti-tank battalion with twelve guns. Separate anti-tank regiments were equipped with between four and five batteries, each containing four guns.

The 53-K was also exported. Several examples were delivered to Republican forces during the Spanish Civil War, where they were employed by the International Brigades, most notably by the British Anti-Tank Battery of 5 International Brigade, beginning with the Battle of Brunete in July 1937.

At the outset of Operation Barbarossa in 1941, hundreds of 53-Ks were either destroyed in combat, captured or simply abandoned during the rapid German advance. The *Wehrmacht* quickly pressed these captured guns into service, re-designating them as the 4.5cm *Panzerabwehrkanone 184(r)*.

Finland

The Finns attempted to purchase *Pak* guns from Germany in the autumn of 1939, but these efforts were thwarted by the Soviets, who demanded strict adherence to the obligations outlined in the Molotov–Ribbentrop Pact signed by Nazi Germany.

Nevertheless, with the onset of the Interim Peace following the Winter War of 1939–1940, the situation changed, and the first batch of fifty guns was delivered from Germany, possibly via Hungary or Spain, according to some sources. The origin of these initial fifty guns remains somewhat uncertain, as the Germans likely attempted to keep this information secret from the Soviets. Regardless of their source, these *Pak* guns arrived within a few weeks of the end of the Winter War, in late March to early April 1940.

The second batch, comprising 100 guns, arrived directly from Germany in September 1940. The third and final delivery, consisting of fifty guns, reached Finland from Germany in June 1941.

While the Germans did not distinguish between different versions of their *Pak* guns, the Finns categorized them differently into two models: PstK/37 and PstK/40. The primary distinguishing factor in Finnish use was the structure of the gun barrel, which underwent changes by the Germans to simplify manufacturing. The barrel of the K/37 model exhibited a noticeable change in thickness around the middle length, whereas the K/40 model had a consistent angle of thickness throughout its length.

The Finns procured a large number of spare parts for these guns, which were interchangeable between the two models. Consequently, wartime repairs often resulted in hybrid versions referred to as PstK/37-40 and

PstK/40-37. The guns delivered to Finland were typically equipped with a primary optical sight and a collimator-type backup sight.

The Finnish ammunition inventory for these guns did not include the *Pz.Gr. 40* round, but in response to tank developments, the Finnish Army acquired 5,000 *Stielgranate 41* grenades.

The Finns continued to use these German *Pak* guns until the end of the Second World War, assigning at least one anti-tank company armed with *Stielgranate 41* grenades to every division. With the conclusion of the Continuation War in September 1944, these guns ceased their service as anti-tank weapons. About 140 of them survived the Second World War and remained in storage for potential use until the late 1970s. During this period, some were used for training purposes as infantry guns. Between 1982 and 1984, most of them were sold and exported.

Being short on modern anti-tank guns during the war, the Finns employed up to 200 captured Soviet 53-Ks from the Winter War, designating them as the 45 PstK/37 (45mm *panssarintorjuntakanuuna m/37*, or 45mm anti-tank gun model 1937).

Italy

The first encounters of *Regio Esercito* ('Royal Italian Army') units with *Tak* guns occurred during the fighting in Ethiopia in 1936, where several of these types of guns – acquired from Germany – were captured, albeit with limited ammunition supplies. These guns were subsequently used during the 1940–1941 battles against British forces in Italian East Africa.

In 1938, Germany transferred forty *Pak* guns to the Italian expeditionary corps in Spain, significantly enhancing the Italian units' anti-tank defence capabilities against attacks from Republican tanks. Some of these guns were towed by Italian L3 light tanks (tankettes). By the spring of 1939, with the end of the Spanish Civil War, these guns were left at the disposal of General Franco's army.

Regio Esercito units received German anti-tank guns once again in 1940, when the Italian Army acquired a batch of 100 *Paks,* which arrived in Naples in December 1940. Of these guns, eighty-seven were sent to Libya and the remainder were held in Italy for possible shipment to Italian East Africa. The gun saw service in North Africa, particularly during the Axis offensive of spring 1941, when it was assigned as part of the anti-tank assets to the *Ariete* Armoured Division, as well as to the *Bologna*, *Brescia* and *Pavia* divisions.

In the desert, the Italians sometimes mounted *Pak* guns on lorries and fired them from these mountings, but this was not a common practice.

By 1942, these guns were being phased out in favour of the heavier Italian 47/32.

In 1943, around a dozen of these guns fell into German hands after the Italian surrender to the Allies. They were used by some second-line German units as the *Pak-162 (it)*; some were installed in the fortified defensive lines across Italy.

Spain

During the Spanish Civil War, *Panzergruppe Drohne*, the anti-tank units of the Condor Legion, trained the Francoists in the use of anti-tank guns at Las Arguijuelas Castle (Cáceres), as such units did not exist in Spain before the outbreak of the conflict.

In the first battles involving tanks (Madrid and Alcázar de Toledo), infantry support guns that could fire armour-piercing projectiles were used against armoured fighting vehicles. Along with the arrival of tanks for both sides (Italian and German for the Nationalists, and Soviet for the Republicans), anti-tank guns also arrived. The first twenty-four 3.7cm *Pak* guns of *Panzergruppe Drohne* were shipped in October 1936 and were swiftly employed in combat. These were followed by 290 further *Pak* guns acquired by the Francoists, allowing for the creation of twenty-eight anti-tank batteries and several sections composed of between three and five guns.

The *Pak* guns proved to be effective against the tanks of the Republican army, including modern Soviet T-26s, as well as against strongpoints.

Many experts, including British and Soviet, believed that during the Civil War, anti-tank weapons had surpassed tank development. They argued that light anti-tank weapons had the advantage of being easily shifted from location to location and quickly brought up to the front lines.

The Republican army was supplied by the Soviet Union with a larger-calibre version of the *Pak*, the 53-K (45mm). The first fifteen 53-Ks arrived on 29 April 1937 in Cartagena, followed by 100 more units unloaded at Cabo de Palos on 7 May. With these initial deliveries, thirty-seven batteries of three guns each were organized, becoming the accompaniment batteries of the recently formed mixed brigades of the Republican army. Many of these guns, when captured, were also used by the Francoists in their anti-tank units.

Hungary

In 1935, Hungary purchased around twelve 3.7cm *Tak* guns. Later, the Hungarian Army received or purchased approximately 300 3.7cm *Pak* (*36M paceltóró agyu*) guns, and produced 40mm *40M paceltóró agyu* guns, utilizing the lower carriage and elements of the upper carriage (including armoured shield) from the German 3.7cm *Pak* guns.

Estonia

In 1937 and 1938, Estonia purchased forty-four 3.7cm *Pak* guns (known in Estonian as the 37mm *M1937 Tankitorjekahur*), with a commercial version featuring a longer *L/50* barrel. These guns were deployed in three anti-tank companies within infantry divisions of the Estonian Army.

Greece

In 1939, Greece purchased twenty-four 3.7cm *Pak* guns and 43,020 rounds of ammunition, including 35,000 rounds of 3.7cm *Pz.Gr.*, 7,200 rounds of 3.7cm *Sp.Gr. 18* and 1,200 training rounds. Between 1939 and 1940, Greece unsuccessfully tried to purchase an additional thirty-six anti-tank guns. The German guns were placed in fixed fortifications on the Metaxas Line, being used in April 1941 to combat attacking *Wehrmacht* and *Waffen-SS* units.

China

In 1930, a license for the production of 3.7cm *Tak* guns was sold to China, where they were manufactured – or more likely assembled from parts provided by the Germans – at an artillery factory in Changsha under the designation Type 30. Given that the Japanese Army fighting in China did not have tanks with anti-shell armour, 3.7cm guns were a very effective means of anti-tank defence for the Chinese.

After the mass production of *Pak* guns was established in Germany, Rheinmetall exported the 35/36 model to China with the standard *L/45* (1,665mm) barrel length, as well as a commercial version with a longer *L/50* (1,850mm) barrel. The recipient of both types of 3.7cm guns was the *Kuomintang* of Nationalist leader Chiang Kai-shek, which, under a contract with the Soviet Union executed in June 1938, additionally acquired, among other types of weapons, eighty Soviet 1-K 37mm anti-tank guns, a Russian licensed copy of the *Tak/Pak*.

Japan

During the battles in China in the late 1930s, the Japanese Army captured a significant number of *Tak/Pak* guns. From November to December 1937 alone, around 100 of these guns were retrieved from the battlefields in the Nanking operation area. After thorough testing, the Japanese adopted the German anti-tank guns for the Imperial Japanese Army, designating them as Type *Ra* 37mm (with *Ra* being a Japanese abbreviation for the

Rheinmetall company). It is important to note that the guns used by the Japanese were exclusively captured equipment, not purchased or produced under license.

During testing, it was observed that the ammunition for the German guns was interchangeable with that of the Japanese Type 94 anti-tank guns. Before distributing the new guns to front-line units, they were modified into three variants designated as *Ko*, *Otsu* and *Hei*. The *Ko* variant was essentially a slightly modified German anti-tank gun. *Otsu* featured a new Japanese chassis adapted for horse-drawn transport, while *Hei* used steel wheels from the Japanese Type 94 gun (this was the least common variant). Some Japanese *Paks* were equipped with a small castoring wheel under the joined trails for easier manhandling, and the shield was often removed to lighten the gun for jungle transport.

The Japanese likely had at least 350–450 operational *Ra* 37mm anti-tank guns. Most of these were later used in combat against Allied forces in the Pacific. For example, in February 1942, the Japanese Sixteenth Army attacking the Dutch East Indies had 169 *Ra* 37mm guns and only seventy Japanese Type 94 anti-tank guns. The captured *Ra* guns remained in service with Japanese forces until the end of the war, and were used during the battles on Guadalcanal in 1944.

Ethiopia

In 1935, Germany, still resentful over Italy's defence of Austria against a Nazi takeover in 1934–1935, sold a dozen 3.7cm *Pak* guns to Ethiopia, along with an undisclosed number of rounds. These weapons were intended to bolster Ethiopia's defences. However, during the course of the Italo–Ethiopian War, Italy subsequently captured these guns.

Romania

During 1943 and 1944, small quantities of 3.7cm *Pak* guns reached the anti-tank units of the Romanian Army. This proved particularly disastrous during the Soviet encirclement (Operation Uranus) at the Battle of Stalingrad, when the already understrength Romanian Third and Fourth Armies bore the brunt of the main Soviet armoured thrust and were unable to stop the Soviet advance due to their grossly inadequate anti-tank weaponry.

Other

Unknown numbers of *Pak* guns were provided for the Slovakian Army, while captured guns were also used by the British Army in Africa and by the Soviet Red Army.

Red Army

Soviet anti-tank gunners manoeuvre their 45mm M1937 (53-K) gun into firing position. Designed in 1937 as an enlarged development of the German 3.7cm *Pak*, this gun adopted a 45mm calibre to take advantage of large reserves of captured French 47mm ammunition, which could be converted by machining down the driving bands.

A Soviet anti-tank platoon equipped with 45mm 53-K guns, towed by T-20 *Komsomolets* artillery tractors, photographed in November 1941 near Samara. (Photo by Mark Borisovich Markov-Grinberg)

While the German 3.7cm *Pak* had been relegated to secondary units by 1944, the Soviet 53-K was still widely employed in front-line formations. This photograph shows a 53-K in action near a bridge over the Dvina during the liberation of Vitebsk in the summer of 1944. (Photo by Samariy Gurariy)

A platoon of T-20 tractors and 53-K anti-tank guns, apparently abandoned rather than destroyed. The weapons appear largely intact, though their rubber tyres are missing – likely taken as 'war souvenirs' by local villagers.

A Soviet 53-K anti-tank gun in German hands, with whom it was re-designated the *Pak-184(r)* ('r' standing for *russisch*, or 'Russian').

Since the *Pak 184(r)* was very similar to the German 3.7cm *Pak*, training requirements were minimal, which allowed for its rapid and widespread employment. According to *Wehrmacht* records, 849 pieces of the *Pak 184(r)* were officially listed in German inventories in 1942. These guns continued to see service into 1943, particularly on the Eastern Front and in rear-area defensive roles, but were gradually phased out in favour of more powerful German and captured weapons, such as the *Pak 38*, *Pak 40* and 7.62cm *Pak 36(r)*.

(**Above**) A *Pak* crew from the *SS-Gebirgs-Division 'Nord'* towing their gun near Tall on Finland's Karelian Front in September 1941, after successfully knocking out several Soviet BT tanks. Due to their thin 13mm armor, these tanks were unable to withstand even the obsolete German anti-tank guns.

(**Middle**) Soviet soldiers examine a captured 3.7cm *Pak* gun, a weapon closely resembling their own 53-K anti-tank gun. While some German *Pak* guns were captured and pressed into Soviet service, their use remained limited due to the lack of mass reserves of compatible 3.7cm ammunition.

(**Left**) Red Army soldiers move a captured 3.7cm *Pak* gun on the Leningrad Front in December 1941.

(**Opposite, above**) Finnish gunners struggle to heave their 3.7cm *Pak* across the uneven cobblestones near Troitsa in 1941. With no motor transport at hand, sheer muscle power becomes the only means of moving the weapon forward. (SA-kuva)

(**Opposite, below**) A Finnish anti-tank gun crew braces their weapon in a hastily dug position at Sirkiänsaari in January 1944, as Soviet armour grinds forward through the snowbound battlefield. The gun is emplaced deep in the drifts, its barrel jutting from a white shroud of winter, while the crew – clad in heavy white overcoats – stand out like phantoms against the frozen landscape. (SA-kuva)

(**Above**) The Second Italo–Ethiopian War (1935–1936) was part of the Abyssinian Crisis, during which Italian armed forces invaded and occupied Ethiopia. Italy had a significant technological advantage over the Ethiopians, including the use of tanks and bombers, which their opponents did not possess. Note that the Italian gun crew are wearing the distinctive pith helmet.

(**Opposite, above**) The Italian crew of a 3.7 cm *Pak* gun stand ready during summer training exercises in Italy. Clad in steel helmets that glint under the Mediterranean sun, the soldiers handle their weapon with precision and discipline.

(**Opposite, below**) Italian forces launch an assault on a British position in Marmarica, Libya, during the autumn of 1941, with a *Cannone Controcarro da 37/45* providing vital support. The gunners forego helmets for the distinctive *Corazzato* headgear, while the upper portion of the gun's shield is lowered to maximize visibility, exposing them to enemy fire.

(**Above**) The Nationalist Spanish crew of a 3.7cm *Pak* on the move to the front line. It small size, lightness and high muzzle velocity made this gun a versatile and effective weapon. The gun's calibre was sufficient to take on armoured vehicles as well as strongpoints.

(**Opposite, above**) Two civilian engineers adjust the sighting telescope on a 3.7cm *Pak* gun. (Bundesarchiv Bild 146-1968-049-04A)

(**Opposite, below**) Under the supervision of German instructors from *Panzergruppe Drohne*, Spanish recruits were trained in the use of *Pak* guns at the Tank School in Cubas de la Sagra. The trainees proved to be highly capable, as the majority of Republican (Soviet-supplied) tanks during the Spanish Civil War fell victim to German anti-tank guns. (Bundesarchiv Bild 146-1974-127-11)

(**Above**) A mixed anti-tank gun crew, combining Spanish trainees in sidecaps and German instructors in black berets, manoeuvre their Pak gun toward a firing position.

(**Opposite, above**) Members of 15 International Brigade pose with a Soviet-made 45mm light anti-tank gun. The first of these weapons entered Republican service during the Battle of Brunete.

(**Opposite, below**) Captured Soviet T-26 tanks, armed with 45mm 20K guns, operated by the Nationalist Armoured Regiment. Based on tactical markings, these vehicles belonged to the 2nd Battalion's 1st Company. To prevent confusion with Republican T-26 tanks, the Nationalists painted red-yellow-red stripes on the turret front and sides and added a St Andrew's cross on a white background atop the turret as an air recognition mark. This photograph was taken during the Battle of the Ebro in 1938. (BNE)

135

(**Above**) A Hungarian 3.7cm *Tak* gun engaging Slovak forces at Rahó (Rakhiv), Subcarpathia, in March 1939. Note that two crew members are holding down the trail spades, as the gun is positioned on a paved road where they cannot be dug in. The clash between Slovakia and Hungary lasted only 100 hours, during which one of the *Tak/Pak* guns knocked out a Czechoslovak LT vz. 35 light tank.

(**Opposite, above**) Soldiers of the Hungarian 24th Border Guard Battalion in front of Khust, a city on the Khustets River, during the Slovak–Hungarian clash of March 1939. The troops are shown guarding the vital river crossing with a 3.7cm gun, a snapshot of the brief but intense struggle over Carpathian territory.

(**Opposite, below**) The blocking position of Hungarian motorized rifle troops on the outskirts of the village of Kirmasovka in the Soviet Union on 24 July 1941 consisted of a 3.7cm 36M anti-tank gun, covered by a rifle squad armed with a 31M light machine gun and 35M rifles. (HIM)

(**Above**) The Chinese crew of a *Pak* gun during official presentation. The license produced gun was described as highly accurate and was well liked by its crews.

(**Opposite, above**) Soldiers of the elite Ethiopian Imperial Guard artillery units, identifiable by their black collar tabs, during field training with German-supplied Pak anti-tank guns. These weapons were capable of penetrating the armour of Italian CV3/33 tankettes at ranges exceeding 500 metres. *(University of Florida, Special and Area Studies Collections)*

(**Opposite, below**) Italian soldiers inspect captured Ethiopian guns after the Battle of Maychew (Italian: *Mai Ceu*) in March 1936. Among the weapons are Pak guns and a 75mm field gun visible in the centre. The determined assault launched by Ethiopian forces under Emperor Haile Selassie against well-prepared Italian defensive positions failed, forcing the Ethiopians to withdraw and abandon much of their heavy equipment. (AC)

Chapter Nine

Next in Line

5cm *Pak 38*

As early as 1936, German Army planners had anticipated that anti-tank guns with calibres greater than 3.7cm would soon be necessary. In response, Rheinmetall-Borsig was commissioned to develop a 5cm anti-tank gun, which was designated the *Pak 38* (*L/60*). The new gun was completed by Rheinmetall-Borsig in 1939.

The *Pak 38* featured a monobloc barrel threaded at the muzzle to accommodate a small two-baffle muzzle brake. It employed a semi-automatic breech mechanism with a horizontal sliding block, and its recoil recuperator system was hydropneumatic. The carriage was constructed of welded steel and mounted on metal disc wheels with solid rubber tyres. The gun was supported in transit by torsion springs, which automatically locked into position when the tubular split trails were deployed.

For crew protection, the *Pak 38* was equipped with a 5mm armoured shield and apron, including a sighting port on the left side of the shield. A small auxiliary wheel could be attached to the end of the trails to assist in manually manoeuvring the gun into firing position.

With armour-piercing ammunition, the 5cm *Pak 38* was capable of penetrating up to 120mm of armour at 500 metres, sufficient to defeat nearly all Allied tanks of the period.

The *Pak 38* was intended to replace the 3.7cm *Pak* as the standard anti-tank gun in divisional anti-tank battalions and regimental anti-tank companies. However, only two guns were completed in 1939, with an additional 388 produced during 1940. Full-scale production ramped up later, and by the time it ceased in 1943, a total of 9,568 *Pak 38s* had been delivered. Many remained in service until the end of the war.

In addition to the towed version, several variants were developed, including turret-mounted models for tanks (some of which were later repurposed as pintle-mounted coastal defence guns along the Atlantic Wall), an automatic aircraft cannon variant known as the *BK 5*, and the 5cm *Flak 214*, a rare anti-aircraft version.

7.5 cm *Pak 40*

Following the same line of development that led to the 5cm *Pak 38*, by 1939 German ordnance planners were already calling for a more powerful anti-tank weapon. The result was the 7.5cm *Pak 40* (*L/46*), which would go on to become the standard German anti-tank gun during the later years of the Second World War.

Once again, a Rheinmetall-Borsig design was selected for production over a competing Krupp submission. To expedite development, the Rheinmetall-Borsig proposal was essentially a scaled-up version of the *Pak 38*, although it incorporated several important modifications.

The monobloc barrel was fitted with a two-baffle muzzle brake, and the horizontal sliding breech block operated semi-automatically. The recoil and counter-recoil system utilized a hydraulic buffer in conjunction with a hydropneumatic recuperator. For crew protection, the gun was equipped with a 5mm-thick, angular, spaced armour shield. The *Pak 40* was fitted with a direct-fire telescopic sight, an indirect-fire sight and an auxiliary sight, allowing for versatility in target engagement.

The carriage was constructed from welded steel, featuring tubular split trails and light alloy wheels with solid rubber tyres. The gun was equipped with two braking systems: air brakes (operated via the towing vehicle) and manual hand brakes (used to stabilize the gun during emplacement). It could be towed at speeds up to 40km/h (25mph) under suitable conditions.

Visually, the *Pak 40* was distinguishable from its predecessor by its straight, flat shield plates, which replaced the curved armour of the *Pak 38*. Another significant but less visible change was in construction materials: while the *Pak 38* used light alloys, the *Pak 40* relied more heavily on steel. This shift was driven by growing shortages of critical raw materials in Germany as early as 1940, with most available light alloys reserved for aircraft production.

As a result of this change, the *Pak 40* was significantly heavier – weighing approximately 1,500kg in travelling configuration – placing it closer in weight class to light field artillery than earlier anti-tank weapons. This increased mass necessitated larger gun crews and standard motorized towing, placing additional logistical strain on the already stretched German war economy.

The first *Pak 40* was delivered for testing in November 1940, though full-scale production did not begin until 1942. Despite an impressive total output – 23,303 units by April 1945, including 11,728 manufactured in 1944 alone – demand continually outpaced supply. Losses in the field were as high as production, with *Pak 40*s frequently deployed on the front lines, where they were destroyed in combat or abandoned during rapid withdrawals. The gun's considerable weight meant it was difficult to move without a functioning prime mover. If the towing vehicle was destroyed or broke down, the gun was often immobilized and left behind, as the crew could not move it any significant distance on their own.

8.8cm *Pak 43*

Whenever the subject of anti-tank guns arises, the Krupp-designed 8.8cm *Flak 18* invariably comes up. Although originally developed as an anti-aircraft gun, it had already proven effective against armoured vehicles during the Spanish Civil War, where it was pressed into service as a stopgap anti-tank weapon, with devastating results. Its capabilities were further demonstrated during the 1940 campaign in France, where it was employed against both tanks and fortifications. By that time, solid armour-piercing projectiles for ground-to-ground use had already been developed.

Despite the widespread battlefield use of the 8.8cm *Flak 18/36/37* guns in the anti-tank role, it is important to note that these were far from ideal solutions. They were cumbersome to deploy, highly visible and difficult to manoeuvre. Nonetheless, it took time for a purpose-built anti-tank version to reach production.

The specialized anti-tank version was eventually designated the 8.8cm *Pak 43*, widely regarded by military historians as one of the finest anti-tank guns of the Second World War. It featured a 71-calibre barrel mounted on a low-profile cruciform carriage that allowed for 360° traverse, protected by a sloped armour shield. Firing the *Pz.Gr. 39/43* armour-piercing projectile, which weighed 7.3kg, at a muzzle velocity of 1,130m/s, the *Pak 43* could penetrate up to 159mm of armour at 2,500 metres – more than enough to defeat any Allied tank of the period.

The tank-mounted version of this gun became the 8.8cm *Kwk 43*, the main armament of the formidable Tiger II heavy tank.

However, this outstanding performance came at a cost. The *Pak 43* was complex and resource-intensive to produce, requiring significant amounts of high-grade steel and manufacturing precision at a time when the German war economy was already stretched thin. Although production centres were established by Henschel in Kassel and Weserhütte in Bad Oeynhausen, early output was minimal: only six units were completed by the end of 1943. In 1944, total production rose to 1,766, still far below the nearly 12,000 *Pak 40*s produced that same year.

During development, it became clear that a simplified carriage would be necessary to accelerate deployment. The result, born of wartime improvisation, was the *Pak 43/41* – a hybrid anti-tank gun using the split-trail carriage of the 10.5cm *leFH 18* field howitzer, combined with the wheels of the 15cm *sFH 18*. A new armour shield was added, and the gun was configured to allow limited indirect fire if needed. To speed production, the breech mechanism reverted to the more familiar horizontal sliding block, compatible with existing manufacturing lines.

Though ballistically identical to the *Pak 43*, the *Pak 43/41* was considerably more awkward and heavy, weighing nearly 1,000kg more. Its limited traverse, high profile and cumbersome mobility earned it nicknames like *Scheunentor* ('barn door') among German crews. Nonetheless, its power was desperately needed on all fronts as Allied tanks were continually upgraded with thicker and better-sloped armour.

The *Pak 43/41* helped fill the gap, but numbers were never sufficient. Production in 1943 totalled 1,152 units, with just 251 more produced in 1944 before manufacturing ceased.

From Wheels to Tracks: the Evolution of Gun Mobility

As successive models of anti-tank guns were developed to replace the relatively light 3.7cm *Pak*, each became significantly heavier and required increasingly robust towing solutions. This shift made it necessary to transition from traditional wheeled towing vehicles to tracked and half-tracked tractors, particularly after the onset of the Russian campaign. There, wheeled transport proved woefully inadequate in the infamous *Rasputitsa* – the seasonal mud that rendered roads impassable.

The *Sd.Kfz. 10*, a light half-track tractor, was the standard towing vehicle for the 5cm *Pak 38* and its equivalents. It also served as a personnel and limber vehicle, offering limited cross-country capability. However, as larger guns such as the 7.5cm *Pak 40* entered service, the need for more powerful towing platforms became evident. The *Sd.Kfz. 11*, a medium half-track, was designated as the standard tractor for this weapon.

Yet in practice, any vehicle with equivalent towing capacity could be employed. Among the most common substitutes were the *Raupenschlepper Ost* – a fully tracked logistical vehicle designed for Eastern Front conditions – and the *Maultier* ('Mule'), a hybrid half-track converted from standard 3-ton trucks by replacing the rear axle with a tracked suspension.

For the massive 8.8cm *Pak 43/41*, introduced later in the war, the standard towing vehicle was the *Sd.Kfz. 7*, a heavy prime mover.

While these tracked and half-tracked tractors did improve mobility, especially off-road, they could not overcome the severe limitations posed by the deep mud and vast distances of the Eastern Front. Towed guns, regardless of tractor type, remained vulnerable to weather and terrain.

This operational reality led to a pivotal shift in doctrine and design: the need for self-propelled anti-tank guns. Mounting anti-tank guns directly on mobile platforms was the only way to ensure both tactical mobility and sufficient firepower.

The Rise of Self-Propelled Anti-Tank Guns

Early efforts to field self-propelled anti-tank weapons were mostly improvisational. These vehicles often used modified – or even unmodified – tank chassis fitted with an open-topped, lightly armoured superstructure. The armament typically consisted of standard anti-tank guns repurposed for vehicle mounting. In many cases, captured equipment from Czech, French or Soviet stockpiles was also adapted to fill the urgent demand.

Though makeshift, these first-generation vehicles significantly increased battlefield mobility and allowed anti-tank units to keep pace with mechanized formations. Many were mounted on the chassis of obsolete or damaged tanks, providing a practical use for otherwise sidelined vehicles.

A turning point came in 1942, when the powerful long-barreled 7.5cm gun was mounted on the *Sturmgeschütz III* (*StuG. III*) chassis. Originally designed as an assault gun for infantry support, the up-gunned *StuG* quickly proved itself to be an exceptional tank destroyer. With its low silhouette, armoured protection and mobility, the *StuG. III* offered a highly survivable and effective platform for anti-tank warfare.

This success led to the development of a new class of purpose-built self-propelled anti-tank vehicles known as *Jagdpanzer* (tank hunters). These combined the roles of tank destroyer and assault gun in a single chassis. Heavily armed and armoured, with sloped frontal armour and low profiles, *Jagdpanzers* were well-suited to defensive warfare and ambush tactics.

By the final years of the war, production of self-propelled guns surpassed that of traditional tanks, reflecting the growing German emphasis on defensive operations and the high effectiveness of these specialized vehicles.

Side by side somewhere in Russia: a 5cm *Pak 38* stands poised in front of its predecessor, the 3.7cm *Pak*. The contrast between the two reflects the rapid evolution of German anti-tank weaponry, as the lightly armed 'door knocker' gave way to the more potent and capable anti-tank gun.

Static firing drill, with 3.7cm *Pak* guns in the background and a 5cm *Pak* in the foreground. The crews reinforced their firing positions with stones to improve stability, maximizing accuracy and control during live-fire exercises.

The *Pak* crew manoeuvred their guns through rugged mountainous terrain, deliberately seeking elevated positions above the valleys. From these commanding vantage points, they could control and dominate the approaches used by motorized columns, turning natural heights into strategic artillery advantages.

Under the relentless sun somewhere on the Eastern Front, German soldiers toil over the maintenance of their *Pak* gun. Sweat gleams on their backs as they tighten, adjust and clean, each movement a blend of fatigue and precision. The gun, a lifeline against advancing Soviet armour, stands ready only through their relentless labour.

A *Pak 38* crew from *Pz.Jgd.Kp. 14*, attached to an unidentified German infantry regiment, photographed in Chrystyniwka near Vinnytsia during the summer of 1941. As infantry divisions were among the last to receive proper motorized tractors, they relied on mobilized civilian tracked vehicles to tow their guns.

A *Pak 38* crew braces their weapon for a fire mission against an unseen Soviet target during the summer of 1942. In the background, the white huts of a Ukrainian village – silent, abandoned and empty, a ghostly witness to the violence about to unfold.

Germans captured roughly 2,000 Renault UE light armoured tractors from France. Before Operation Barbarossa, many were dispatched to anti-tank companies, serving alongside 3.7cm and 5cm guns as nimble gun tractors and ammunition carriers.

The 5cm gun, significantly heavier than its 3.7cm counterpart, was nearly impossible to transport with small dinghies. To move these formidable weapons across rivers and streams, crews relied on larger boats specifically for the task, carefully balancing the weight as they navigated the water.

Tank hunter personnel of the *97. Jäger Division*, which fought on the Eastern Front from 22 June 1941, initially in Ukraine (around Lviv, Vinnytsia, Uman and Kyiv) and later on the Donets River.

The crew of a 5cm *Pak 38* pose for the camera during a pause in the fighting on the Eastern Front during the summer of 1942.

(**Above**) Towed *Pak* guns were vulnerable to carriage damage, particularly when traversing the rough and often poorly maintained roads of the Eastern Front. As a result, gun crews preparing for long marches routinely carried essential spare parts for emergency field repairs, including critical components like main wheels.

(**Opposite, above**) The crew of a 5cm *Pak 38* gun prepares their weapon for a fire mission against a Soviet target. These guns were normally assigned to divisional anti-tank battalions, but were often seen supporting grenadier anti-tank companies.

(**Opposite, below**) An anti-tank crew at work carrying out routine maintenance, clad in training uniforms, concentrated amid the barracks courtyard – a moment of preparation that underscores the unseen labour behind battlefield readiness.

Surrounding their 5cm *Pak 38*, the crew stands at ease, helmets gleaming and uniforms taut, a frozen moment of resolve before the chaos of the battlefield.

Photograph of *Luftwaffe* officers manoeuvring a 5cm *Pak* gun, possibly as part of a friendly bet to test their strength and endurance – a brief, spirited moment of challenge and camaraderie captured in time. Note that the gun is painted in overall *Dunkelgelb* (dark yellow), which was adopted as standard camouflage colour in February 1943.

A *Pak 38* gun crew conducting a field check after completing maintenance, carefully inspecting their weapon to ensure it is fully operational.

An intriguing photograph showing a heavy gun tractor towing both an 8.8cm and a 3.7cm *Pak* gun. The stark contrast in their weights is evident: the larger gun tipping the scales at around 4 tons, while the lighter 3.7cm barely exceeds 300kg.

(**Above**) The comparison between the late-war 8.8cm *Pak 43/41* anti-tank gun and the early-war 3.7cm *Pak* gun highlights a significant evolution in anti-tank weaponry during the Second World War, primarily evident in their size. The differences in barrel length, calibre and projectile velocity allowed the 8.8cm to more than triple its effective range versus the 3.7cm, extending its ability to engage heavy armoured targets at a distance.

(**Opposite, above**) The light half-track *le. Zgkw. 3t (Sd.Kfz. 11)* hauling a 7.5cm *Pak 40* provided seating for the driver, the commander and six crew members. Ammunition was stored in the large stowage box in the centre of the superstructure.

(**Opposite, below**) An *Sd.Kfz. 10* half-track retreats past burned-out barns, illustrating the scorched earth tactics employed by German forces during withdrawals.

An *Sd.Kfz. 10* with a mounted 5cm *Pak 38* takes a short break on the front lines in 1943. By this stage of the war, German forces were up-gunning, converting and modifying their anti-tank weapons to increase mobility and firepower in response to the growing threat of heavily armoured enemy tanks.

By 1942, some Renault tractors remained in service with German anti-tank units, primarily used to tow 3.7cm and also some 5cm *Pak* guns.

A German battlegroup, consisting of a tank platoon and an anti-tank half-battery, advances toward a village on the Eastern Front. Rather than moving along the main road, which might have been mined or ambushed, the tanks manoeuvre around the huts to flank potential defensive positions.

A German *Sd.Kfz. 10* tows a 5cm *Pak 38* gun through a narrow street in a Soviet city during the summer of 1941. The column passes a Russian sign reading '*БЕНЗИН*' ('petrol'). Since knowledge of Cyrillic was very limited among German troops, *Wehrmacht* traffic control units supplemented such signs with the Latin letter 'B' for *Benzin* ('fuel').

Bibliography

Books

Fleischer, W. & Eiermann, R., *Die deutsche Panzerjagertruppe 1935–1945.*

Haupt, W., *3.7cm Panzerabwehrkanone (Pak)*, Waffen Arsenal Band 29, 169.

Hogg, L., *German Artillery of World War Two* (London, 1997).

Koch, M., *Merkblatt fur die Munition der 4,2 cm Panzerjagerkanone 41 (4,2 cm Pak 41)* (Berlin, 1942).

Kosar, R., *Panzerabwehrkanonen 1916–1977* (Stuttgart, 1978).

Pawlas, K., *Waffen-Revue Nr. 72, 73 i 74 (3.7cm Pak)* (Schwabisch Hali, 1977).

Pawlas, K., *Waffen-Revue Nr. 75 (4,2cm 72,8 m Pak 41)* (Schwabisch Hali, 1978).

Pawlas K., *Waffen-Revue Nr. 127, Datenblatter rur Heeres Waffen, Fahrzeuge und Gerat* (Niirnberg, 1976).

von Senger und Etterlin, R., *Die deutschen Geschutze 1939–1945* (Bonn, 1998).

Documents and Regulations

H.Dv 481/74. Uberblick iiber den Rustungsstand des Heeres, Dezember 1943 und Marz 1944.

H.Dv 481/73 Merkblatt fur die Munition der 3.7 cm Panzerabwehrkanone (3.7 cm Pak), 1938.

Enemy weapons. Part III. German light anti-aircraft and anti-tank guns, The War Office, 1942.

H.Dv. 470/6 Anhaltspunkte für die Ausbildung der Panzerabwehreinheiten.

Von Borries; Dienstunterrichtsbuch fur die Panzerabwehr-Kompanie, Berlin, 1933.

Von Cochenhausen, Taktisches Handbuch für den Truppenführer und seine Gehilfen, Berlin, 1936.

WWW

http://www.niehorster.org/

https://www.lexikon-der-wehrmacht.de/

Back Cover Story

Small, light and deadly in its day, the 3.7cm *Pak 35/36* was the German Army's first true standard anti-tank gun. Conceived in the aftermath of the Great War and tested in the Spanish Civil War, this modest weapon became a cornerstone of the *Wehrmacht*'s early campaigns. From the storming of Poland to the sweeping victories in France and the Low Countries, the 3.7cm *Pak* earned a reputation as the spearhead of the *Blitzkrieg*.

But the rapid evolution of armoured warfare soon tested its limits. On the Eastern Front, the little gun faced the steel giants of the Red Army – an unequal contest that forced improvisation and adaptation. Mounted on Panzer III tanks, pressed into service on a host of makeshift self-propelled chassis and deployed in countless field modifications, the 3.7cm *Pak* proved far more versatile and enduring than its nickname 'door knocker' suggests.

This book traces the weapon's design, development and combat career, with detailed coverage of organization, tactical doctrine and operational use. Richly illustrated with rare and previously unpublished photographs, the story offers fresh insights into one of the most iconic weapons of the early war years.

Author's Bio

Jacek Zabielski, a banking expert from Poland with a passion for modern history, holds a Master's degree in International Relations from Warsaw University in Poland and in European Studies from Central European University, Prague, an MBA from National Louis University in the USA, and has completed a postgraduate programme in Political Science and Media Policy at the University of Amsterdam. His historical research has primarily centred around contemporary military technology, with a particular emphasis on Central and Eastern Europe.